twilight
of american
sanity

ALSO BY ALLEN FRANCES

Saving Normal
Essentials of Psychiatric Diagnosis

twilight
of american
sanity

A PSYCHIATRIST ANALYZES THE AGE OF TRUMP

Allen Frances, MD

wm

WILLIAM MORROW
An Imprint of HarperCollins*Publishers*

With belated thanks to Joe Frances—my late (and, in his own humble way, great) father, who taught me how to see through blowhards like Trump and how simple it is to be happy

HarperCollins books may be purchased for educational, business, or sales promotional use. For information, please email the Special Markets Department at SPsales@harpercollins.com.

A hardcover edition of this book was published in 2017 by William Morrow, an imprint of HarperCollins Publishers.

FIRST WILLIAM MORROW PAPERBACK EDITION PUBLISHED 2018.

Library of Congress Cataloging-in-Publication Data has been applied for.

ISBN 978-0-06-239451-4

18 19 20 21 22 LSC 10 9 8 7 6 5 4 3 2 1

The iniquity of the fathers will be visited on the children and the children's children, to the third and the fourth generation.

—EXODUS

As democracy is perfected, the office of the president represents, more and more closely, the inner soul of the people. On some great and glorious day the plain folks of the land will reach their heart's desire at last, and the White House will be adorned by a downright moron.

—H. L. MENCKEN

A human being is a part of the whole, called by us "universe," a part limited in time and space. He experiences himself, his thoughts and feelings as something separated from the rest, a kind of optical delusion of his consciousness. This delusion is a kind of prison for us, restricting us to our personal desires and to affection for a few persons nearest to us. Our task must be to free ourselves from this prison by widening our circle of compassion to embrace all living creatures and the whole of nature in its beauty.

—ALBERT EINSTEIN

Contents

Trump Isn't Crazy. We Are.

Insanity in individuals is somewhat rare. But in groups,
parties, nations, and epochs, it is the rule.
— FRIEDRICH NIETZSCHE

I t is comforting to see President Donald Trump as a crazy man, a one-off, an exception—not a reflection on us or our democracy. But in ways I never anticipated, his rise was absolutely predictable and a mirror on our soul.

Early in the recent U.S. presidential campaign, a producer invited me to appear on a national television program to analyze Trump's psychology and provide him with a psychiatric diagnosis. It might have been fun. Trump and I were born around the same time and grew up within a few miles of each other in Queens, outside Manhattan. I had casually followed his lifelong quest for media attention, finding his misadventures and constant self-promotions cute, in a repulsive sort of way.

But I had to decline the invitation. First off, I saw no evidence that Trump had a mental disorder and, even if I did, the American Psychiatric Association has a useful ethics policy that explicitly prohibits the armchair diagnosis of politicians. It goes back to the 1964 presidential election. Liberal psychiatrists had taken cheap shots at Barry Goldwater, the radical right-wing Republican candidate—publicizing their "diagnosis" that he was too mentally ill to be trusted with the nuclear button. They had no right to use professional credentials to medicalize their political beef with Goldwater, and I had no desire to register my political disapproval and personal distaste for Trump via psychiatric name-calling. When, instead, I jokingly offered to express my purely layman's opinion that Trump is a classic "schmuck," the producer quickly replied that this wasn't the least bit newsworthy—everyone already knew it. I laughingly agreed—but, as it turned out, we were both dead wrong.

To everyone's amazement (probably including his own), Trump was elected president of the United States. The blustery, bullying bravado that served him so well on reality TV and on the campaign trail is a disastrous disqualification in America's commander in chief. In his own inimitable way, Trump bragged that, once president, he would become "more Presidential than anybody other than the great Abe Lincoln . . . he was very Presidential, right?" But Trump can never be anything but Trump. We have had our share of dumb presidents, impulsive presidents, lying presidents, ignorant presidents, narcissistic presidents, bellicose presidents, conspiracy theory presidents, and unpredictable presidents—but never before has one president so fully embodied all these reprehensible traits. And never before have the institutions of American democracy appeared so fragile in the face of autocratic attack. Trump has scared so many people that six dystopian classics have suddenly jumped to the top of Amazon's bestsellers list (Orwell's *1984* and *Animal Farm,* Huxley's *Brave New World,* Sinclair Lewis's *It Can't Happen Here,* Margaret Atwood's *Handmaid's Tale,* and Ray Bradbury's *Fahrenheit 451*).

Trump's mental health (or lack thereof) is a trending topic on the Internet; on cable news programs; in magazines and newspapers; and most hilariously on *Saturday Night Live.* And political pundits, politicians, and comedians pored over the so-called Bible of Psychiatry, the *Diagnostic and Statistical Manual of Mental Disorders* (*DSM*), and reached the consensus that Trump suffers from Narcissistic Personality Disorder. Soon many psychologists and some psychiatrists (compelled by a perceived higher calling of national interest) chimed in to break the restriction against diagnosing politicians at a distance. Numerous patriotically worded petitions were initiated. A typical one, garnering more than fifty thousand signatures, declared, "We, the undersigned mental health professionals, believe in our professional judgment that Donald Trump manifests a serious mental illness that renders him psychologically incapable of competently discharging the duties of President of the United States. And we respectfully request he be removed from office according to article 4 of the 25th amendment of the Constitution."

I wrote the criteria for Narcissistic Personality Disorder that first appeared in *DSM-III* and are still in use in *DSM-5,* the most recent edition. Trump's amateur diagnosticians have all made the same fundamental error. They correctly note that the disorder's defining features fit him like a glove (grandiose self-importance; preoccupations with being great; feeling special; having to hang out with special people; requiring constant admiration; feeling entitled; lacking empathy; and being exploitive, envious, and arrogant). But they fail to recognize that being a world-class narcissist doesn't make Trump mentally ill. Crucial to the diagnosis of Narcissistic Personality Disorder is the requirement that the behaviors cause clinically significant distress or impairment. Otherwise, many, if not most, politicians (and almost all celebrities) might qualify. Trump is a man who causes great distress in others but shows no signs himself of experiencing great distress. His behaviors, however outrageous and objectionable, consistently reap him fame, fortune, women, and now political power—he has been generously rewarded for his Trumpism, not impaired by it. Trump is a threat to the United States, and to the world, not because he is clinically mad, but because he is very bad.

I hate it when psychiatric diagnosis is so carelessly misused to mislabel as mental illness every conceivable example of simple bad behavior. I was heavily involved in the preparation of *DSM-III* and in charge of the preparation of *DSM-IV.* This burdens me with the responsibility to keep psychiatric diagnosis as honest and as accurate as possible. Most mass murderers are not mentally ill. Most terrorists are not mentally ill. Most rapists are not mentally ill. Most dictatorial rulers are not mentally ill. Most obnoxious boobs are not mentally ill. Most liars aren't mentally ill. Most conspiracy theorists aren't mentally ill. And there's no evidence that Trump is mentally ill. Trump's boorish manners, vulgar speech, and abusive actions make him a national embarrassment and the worst of all possible role models (perhaps there should be a PG-13 rating on all Trump appearances to protect our children from his bad influence). He diminishes America, reducing its greatness. But none of this makes him mentally ill.

There are three harmful unintended consequences of using psychiatric tools to discredit Trump. First, lumping him with the mentally ill stigmatizes them more than it embarrasses him. Most mentally ill people are well behaved and well meaning, both of which Trump decidedly is not. Second, medicalizing Trump's bad behavior underestimates him and distracts attention from the dangers of his policies. Trump is a political problem, not meat for psychoanalysis. Instead of focusing on Trump's motivations, we must counter his behaviors with political tools. And, third, were Trump to be removed from office, his successors (Pence and Ryan) would probably be much worse—more plausible purveyors of his very dangerous policies.

But what does it say about *us,* that we elected someone so manifestly unfit and unprepared to determine mankind's future? Trump is a symptom of a world in distress,* not its sole cause. Blaming him for all of our troubles misses the deeper, underlying societal sickness that made possible his unlikely ascent. Calling Trump crazy allows us to avoid confronting the craziness in our society—if we want to get sane, we must first gain insight about ourselves. Simply put: Trump isn't crazy, but our society is.

---

*Going back hundreds of years, an inverted flag has served as a universal distress signal. I chose the lovely, lonely photo that adorns this book's cover to symbolize America's extreme distress in the age of Trump.

I started this book two years before there was any thought that Trump would enter its pages. It was, and remains, a study of societal insanity—our inability to respond meaningfully to the increasingly urgent dangers that threaten human survival—overpopulation, global warming, resource depletion, and environmental degradation. The grave risks we faced then are now greatly amplified by Trump's aggressive assault on our collective sanity. Danger has always been man's fate—we humans have mastered fearful existential crises every single day of our two-hundred-thousand-year existence. But previously, the scale of risk was relatively restricted—to the individual, family, tribe, city-state, or nation. Current threats are globalized—our planet has become so small and so interconnected, there is no longer a safe place for any of us (even the richest and most powerful) to hide.

Einstein famously defined insanity as "doing the same thing over and over again and expecting different results." Previous civilizations have all mindlessly followed the same depressing cycle of rapid growth and sudden collapse. The tragic mistakes they made then are precisely the mistakes we are making now. Learning from the past is the only way to ensure that our civilization will survive into the future. Sadly, facing reality does not come to us naturally or automatically. A wise saying from the Talmud sums up a great deal of what's problematic in human psychology: "We don't see things as they are, we see things as we are." The path to mastery can be walked only by those whose eyes are open—inconvenient truths don't disappear when disguised by Trump's convenient lies.

We face this perilous tipping point with a psychological makeup much better adapted to the Paleolithic past than to our rapidly evolving present, or extremely risky future. Evolution endowed us with inborn psychological proclivities that worked quite well for our wandering ancestors, ever unsure if they would make it to the next day or where their next meal was coming from. The world, once very big, with seemingly boundless resources, has become small and stretched well past its limits. Survival back then depended on intuitive, short-term decisions—and greed was mostly good. Selfish survival instincts that worked so well fifty thousand years ago now push us to act self-destructively in a world that requires cooperative planning. We have conquered our external world; the question now is whether we can conquer our internal impulses. Maintaining our unrealistic standard of living risks handing substandard living conditions on to our successors.

My job as a psychiatrist has been to help patients learn from their mistakes—uncover the irrational in their thinking and end their vicious cycles of self-defeating behavior. Maturity in an individual (or a society) requires substituting reasoned thought for wishful thinking and immediate gratification. Analyzing Trump's psychology makes no sense because it is too obvious to be interesting and impervious to cure. We can't expect to change Trump, but we must work to undo the societal delusions that created him. We risk destroying our future unless we face it realistically—replacing instinct with rationality and favoring our altruistic impulses in preference to Trumpian self-serving. The tough love that helps heal suffering patients may, I hope, also help heal our sick society.

By lucky nature, I am an optimistic and happy person who loves the present moment and has enjoyed a satisfying past. Still, I can't help but fear greatly for the future. Not my own personal future—I am fully resigned to having not that much time left and feel grateful for the good times I've already had. But, as I approach the end of my life, it saddens me greatly that my generation has been so profligately indifferent to the needs of our children and our children's children. Daily, we privilege our own happiness at the expense of theirs, threatening the future of our species, while also destroying so many of the wonderful life-forms that evolved along with us on this tiny, but dazzlingly beautiful, pebble of a planet. We have enjoyed a rare historical sweet spot, living in the best of times and best of places. Restless human ingenuity, facilitated by ten millennia of unusually stable climate, has blessed readers of this book with unprecedented prosperity; enviable longevity; an ever-advancing array of technical toys; and surprisingly deep knowledge about how our universe works. Previous good fortune has created complacent expectations that each succeeding generation will have things even better than the one before. But, unless we take the difficult steps to make our world sustainable, our kids will be stuck paying our debts. Mistakes made now may bequeath to our grandchildren the worst of times and worst of places. The status quo is not a likely option—we will improve our world or we will destroy it.

Chapter 1

Confronting the Facts of Life

We met the enemy and they is us.
—POGO

Ignorance is not bliss. What you don't know most certainly can hurt you, often in the least expected and most devastating of ways. Compelling evidence indicates that our world is marching blindly into a perfect storm of irreversible catastrophes. Absent urgent corrective action, it will soon be game over for our civilization, with no do-overs allowed. It is very long past time that we face reality straight on, roll up our sleeves, and find practical solutions to seemingly insoluble existential threats. Instead, we harbor a whole set of societal delusions that perpetuate the fatal fallacy that the best way to deal with dangers is to deny their existence. The Trump presidency will be either the final straw or a last-minute wake-up call. His personal positions are wrongly and ridiculously retrogressive on every single issue crucial to the survival of our species. He and his henchmen are every day making decisions that bring us ever closer to an environmental apocalypse and societal breakdown. We have placed the future of humanity in the hands of someone indifferent to facts, proud of scientific ignorance, and ready to act deceitfully on whim and spite. Any one person is entitled to be dead wrong without being called crazy— but it is crazily delusional for our country to follow this clownish Pied Piper unto perdition.

This chapter will subject our most comforting and dangerous societal delusions to some good, old-fashioned reality testing. We will lift the veil of denial and wishful thinking, exposing the false convictions perverting our policies and practices. Not a pretty picture, but ignoring it risks the survival of our clever, but vulnerable species. Delusions die hard. Ideology, expediency, anger, and fear are powerful protectors of even the most unstable and unsatisfactory status quo. The insight that we are digging a deep hole for ourselves is the first and necessary step toward climbing out of it.

In psychiatry, a delusion is defined as a fixed, false belief that is firmly maintained and resists correction by overwhelming evidence and rational argument. Used as a verb, "to delude" means persuading someone to believe something that is false—what many politicians do for a living much of their time (and Trump seemingly does all the time). We don't label everyone delusional just for believing things that aren't true. It's an inherent part of human nature to create inaccurate explanations that comfort us in the face of life's great uncertainties. Myths from ancient times, when we knew so very little, continue to persist today, despite our now knowing so very much. And we are forever creating new myths to help us confront the discomforts of the present and our fears for the future. Most of us have at least some false beliefs and many of us stick to them in the face of what would seem to be compelling contrary evidence. A third or more of Americans manage still to believe in flying saucers, Bigfoot, angels, extrasensory perception, reincarnation, and astrology, and about an equal number manage not to believe in evolution, the Big Bang, the advanced age of planet Earth, the immensity of the universe, or the value of vaccination.

It was once perfectly reasonable to believe the world was only six thousand years old; less so now that geologists know with scientific precision that it dates to 4.543 billion years ago. Someone who clings to the belief in a very young world is dead wrong, but is not considered delusional because his incorrect belief is so widely shared by so many equally mistaken others and doesn't adversely impact his daily life. An individual is diagnosed delusional only when his wrong beliefs are personal, idiosyncratic, and impairing. Individuals who act in response to delusions usually get into big-time trouble. Successful adaptation to life's challenges requires confronting them realistically. Living in a personal dream world means making serious mistakes in real life. Denying and distorting the facts of life is a self-destructive, last-resort defense when reality seems too painful to bear and too imposing to change. I think it's fair to call our society delusional because we've lost touch with the reality of starkly obvious existential threats. Instead we weave wish-fulfilling, convenient, but dangerous fantasies that global warming is a hoax; that the earth's bounty is endless; and that we don't have to limit our world's population.

I have spent a good deal of my life diagnosing and treating people with all sorts of different delusions. The most common are delusions of persecution—"I have enemies everywhere"; "some external force or person is causing all my troubles"; and "when I fail it is not my fault, someone made me fail." The second most common delusions are grandiose—"I am exceptional"; "I have extraordinary powers"; "I have been given a special mission"; "I can do no wrong"; "nothing I do can fail"; "everything I do is condoned by higher forces"; "if I have to hurt people along the way, they are necessary collateral damage." Least common are erotomanic delusions—the stubborn conviction one is secretly loved when in fact one is ignored or hated. Being delusional, in each instance, means stubbornly and defiantly maintaining the fixed false belief in the face of contrary evidence that any rational person would accept.

Our society suffers from delusions quite like those that occur in individuals—in cause, in content, and in consequences. What causes an individual patient to hold fast to patently absurd beliefs, even when these result in great impairment? Partly brain disorganization and misfiring; partly psychological defense mechanisms that help avoid a painful reality; and partly overwhelming social stresses. Why, similarly, would our society stubbornly hold fast to patently absurd beliefs even when these result in great impairment? Partly because of psychological factors within political leaders and their followers, and partly the daunting magnitude of the social, political, economic, environmental, and resource problems that we must now face. In both individual and societal delusions, there is the same denial of intruding reality and the same rush to replace it with persecutory blaming, grandiose posturing, and a false sense of being admired.

As with individual delusions, societal delusions make us blind to risk, heedless of unintended consequences, passive in the face of crises, and dependent on the misguided faith that the future will take care of itself, without tough decisions in the present. Delusional denial allows us to avoid the painful reality that we have already made quite a mess of our world and must make big sacrifices to clean it up. False beliefs that would cause only limited harm when held by a powerless individual become catastrophic when they allow the world's most powerful country to make disastrously bad decisions. Trump is not delusional because tens of millions of people share his beliefs, but he does enable and embellish the societal delusions that can eventually be the death of us. Denying reality is comforting in the short run, but disastrous in the long. If we don't very soon wake up from our delusional reverie, we will find ourselves living in a world beyond repair.

Despoiling the Environment

Societal Delusion: We don't have to worry about global warming or environmental pollution because God or technology will save the day.

Reality Testing: Life-forms profoundly influence the environment they live in and then in turn are profoundly influenced by the environment they have created. Anaerobic bacteria ruled the world until two billion years ago, when they poisoned themselves with their own waste product, oxygen—which luckily became our ancestors' sustenance. We rule the world now but are poisoning ourselves with our own waste product, carbon dioxide—great for plants, but not so great for the future stability of our climate and our sea levels.

Dramatic changes in earth's climate have been the norm over geological time, not the exception. Our civilization has thrived as much as it has only because the climate of the last ten thousand years has been so unusually stable and forgiving. We are now already quite deep into the process of irrevocably destroying that stability. The history of mankind is littered with dozens of specific cultures that collapsed because of natural swings in climate. Our society is the first to plant the seeds of worldwide destruction, by artificially and mindlessly creating a climate completely incompatible with our current population numbers and the distribution of our people.

The effects of global warming are likely already upon us: year after year of record-setting world mean temperatures, unprecedented droughts, unprecedented storms and flooding, Atlanta colder in January than Anchorage, supposedly once-in-a-hundred-year weather disasters happening every few years. Because weather is so naturally variable, scientists have rightly been very cautious in attributing any one event in any given place to long-term, global climate changes.

But with accumulating data, the strongest possible scientific consensus now conclusively confirms that we are rapidly messing up our world, and that big-time corrections must be made now or it will soon be beyond our power to reverse the havoc we will have created. The scientific data all converge: measures of air and sea carbon dioxide, rising temperatures, melting ice caps and glaciers, and different computer models all predicting drastic future temperature and sea level changes. The only remaining deniers of climate change are a very few eccentric oppositionists, the employees of fossil fuel companies, and the recipients of research grants from them. The only real scientific questions left are how fast, how devastating, and how reversible is the damage likely to be. On this there is considerable disagreement, but without much real difference—even the most optimistic assumptions lead to scary scenarios.[1]

The risk of an imminent environmental catastrophe could not be clearer to anyone with eyes to see, but there are four different groups wearing different but equally dangerous delusional blinders: the technophiles, the corporates, the politicos, and the religious zealots.

The technophiles are the most rational of the four irrational branches of global warming deniers. They are great believers that the wonders of advancing technology will provide a last-minute magical save. After all, Victorian London did not, as some had predicted, drown under the mountains of horse doo being produced by its rapidly expanding carriage trade. The problem was erased effortlessly by the invention of the automobile (first called a "horseless carriage," then regarded as so very clean by comparison to its living predecessor). Perhaps we can have our current cake and eat it too—snazzy new techno solutions will come to the rescue, just in time. Too much carbon spewing forth—no big deal. We will sequester it underground or in the ocean depths or shoot up some sulfur particles to balance its heat-trapping effects in the atmosphere. Maybe so. Maybe not. All the suggested gizmos are untried and all carry the considerable risk of being either ineffective or of causing catastrophic unintended consequences of their own (for example, the terrifying thought of snowball earth if we block out too much solar energy).

In any sane world, we would not rely on fairy-tale dreams of future magical fixes, but instead would take out a practical insurance policy right now, by doing everything reasonably within our power to lower worldwide carbon dioxide production, dramatically and immediately. Every responsible person carries insurance against all sorts of life's unforeseen and highly unlikely, but potentially catastrophic, risks. We don't think twice about buying health insurance, home insurance, car insurance, life insurance, liability insurance, and it goes on and on. From a gambler's perspective, these are all sucker bets. Insurance companies can laugh all the way to the bank by tipping the actuarial odds even slightly in their favor, so that they pay off less in reimbursements than they receive in premiums. We are willing to pay the premium because it spreads the risk across a large group, so that each of us may feel secure. We don't wait to develop an illness before taking out health insurance, or have an accident before getting car insurance, or apply for fire insurance after the fire starts.

The whole meaning of insurance is that we buy it early and without any certainty we will ever need to use it. The risk of immediate disaster is minuscule, but people go on buying insurance because not having it would be devastating if the almost impossible did in fact happen. I very much doubt my house will burn down, but I still insure it. The same foresight should inform our approach to taking the expensive steps now that provide future insurance against global warming. We must pay the price well before we can be absolutely certain how grave are the risks and how far in the future is the tipping point. We can't afford to complacently bet the species on the blind hope that someday we can work out a technological fix. Prevention is much better than cure, especially since there may not be a cure or the cure may be too late to save the day. And we buy life insurance, not because it will help us, but because we care about our children. We should have the same protective attitude toward bequeathing our children a safe environment.

Greedy corporate interests have consistently put profit over people—doing everything in their considerable power to promote phony climate science and discredit the inconvenient truth. Ironically, internal documents prove that, beginning in the 1970s, Exxon scientists were pioneers in understanding that catastrophic global warming was an inevitable consequence of the company's practices. But Exxon's leadership has worked tirelessly and deceptively to bury awareness of the risk that had been unearthed by their own scientists. They have spent billions casting doubt on climate science; electing beholden science-denying politicians who can't see past Election Day; scaring everyone about lost jobs; and propagandizing the public into numb inaction.[2] Executives and shareholders think in terms of quarters and short-term return on investment, not in terms of decades or centuries, and certainly not considering the future of humanity. The basic pitch is "do nothing" until we can prove beyond a shadow of a doubt that we are irrevocably ruining our environment. Of course, this is brutal and ruthless cynicism—by then it will be too late to save ourselves. "*Après moi, le déluge.*" Let the grandchildren take the heat.

More oil and low oil prices are the worst possible things for a world heading heedlessly toward lethal climate change. But we are pleased with them because we value current comfort more than we fear long-term catastrophe. In any sane world, we would discourage consumption of fossil fuels by placing a much higher tax on them, particularly when their price drops. To keep things revenue-neutral and balance the pain, other taxes could be reduced and/or cash payments could be distributed for those most harmed by the rising fuel prices. With a simple policy change, we would save the environment, reduce dependence on foreign oil, and preserve our limited fossil fuel supplies so that they will sustain us longer into the future.

President Obama's commendable efforts to develop a sane environmental policy were thwarted every step of the way by the Republican head of the Senate Environmental Committee, James Inhofe. Inhofe is a world-class science denier, a blindly fundamentalist Christian, and a well-paid lackey of the oil industry. His statements on global warming express the essence of societal delusion: "97 percent of climate scientists saying something doesn't mean anything." "Global warming is the greatest hoax ever imposed on the American people." "Increases in global temperatures may have a beneficial effect on how we live our lives." "My point is God's still up there. The arrogance of people to think that we, human beings, would be able to change what He is doing in the climate is to me outrageous." Inhofe may be a true believer, but his passion is also probably inspired by more worldly rewards—

the millions of dollars bestowed by the oil industry and his special relationship to the Koch brothers.

Under Inhofe and Koch brother tutelage, President Trump has picked a cabinet of voracious energy foxes to guard the precarious environmental henhouse. All have unresolvable financial and political conflicts of interest and all have a stubborn ideological attachment to ignorant climate denial. The former head of Exxon, Rex Tillerson, is now Trump's secretary of state. Ryan Zinke, head of the Department of the Interior and a former congressman from Montana, had a 3 percent favorable voting record from conservationists because he voted down the line for every oil project exploiting federal land. Rick Perry at the Department of Energy has called climate change an unproven scientific theory and actually wanted to eliminate the agency he now leads. Ben Carson, housing secretary, believes that climate is always changing, so why worry about it now. The CIA director (Mike Pompeo), attorney general (Jeff Sessions), homeland security head (John Kelly), health secretary (Tom Price), and commerce secretary (Wilbur Ross) are all also clearly on record as strong climate deniers.

Trump's head of the Environmental Protection Agency, Scott Pruitt, is the last man on earth to be entrusted with our survival. He recently made the egregiously ignorant remark, "I would not agree that CO_2 is a primary contributor to the global warming that we see." He is the poster child of societal delusion, bought and paid for by the energy industry. For many years, as attorney general of Oklahoma, he serially sued the EPA to prevent it from doing its job. Now as its head, he has the opportunity to destroy it from the top down—filling the ranks of the EPA with climate deniers, censoring its website, and muzzling its scientists. Pruitt is committed to doing everything in his power to allow corporations to despoil our environment. The Trump administration budget proposal for the EPA would cut it by an astounding one-third and some GOP congressmen have even suggested abolishing the department.

If you were a sworn enemy of human civilization, intent on finding the best way of destroying it through global warming, you could do no better than arranging to get Trump elected president, Mike Pence vice president, Pruitt as head of the EPA, and other energy industry flunkies running every agency in the U.S. government. This is societal delusion in its purist insanity. Trump's pulling the United States out of the Paris climate agreement is a tragic moment in world history. If we tip global warming past irreversibility, millions of people may perish, and Trump must bear some responsibility for their deaths.

Eventually, dire circumstances will persuade us to take the actions needed to forestall global warming. We will recognize the impossibility of exploiting all of our fossil energy reserves without irreversibly befouling our world and destroying our climate. We will sacrifice current happiness and comfort by dramatically conserving energy and investing heavily in sustainable sources for the future. We will be less selfishly concerned about now and more eager to preserve the world for those who will follow. We will take responsibility for our future instead of placing long-shot bets on God or the technophiles. Enough scorching summers, enough droughts, enough "one-hundred-year storms" happening every year, enough floods, enough melting ice caps, enough sea level rise, and even the most devoted followers of Trump and even the greediest beneficiaries of the oil industry will finally come around and see the light. But by then it may be too late. Our wake-up call may come after the horse is out of the barn, when it will be impossible to salvage our world from environmental disaster. An ounce of prevention now is a lot cheaper and safer than desperately playing catch-up with a pound of cure then. Only a badly delusional society could see it otherwise.

The Population Bomb

Societal Delusion: World population can keep growing without causing drastic resource depletion, irreversible global warming, incessant wars, mass migrations, frequent pandemics, and recurring famines.

Reality Testing: Population control has become the most politically incorrect of topics. Even though overpopulation is now directly responsible for virtually every catastrophic problem in our world, the taboo against discussing or dealing with it is almost absolute in the media, political debate, and scholarly presentations. Analyses of the causes of the latest war, refugee crisis, famine, or pandemic almost always stay focused only on the political or economic or personal precipitants—almost never alluding to the most compelling root cause in overpopulation. Everyone is scared off by the admittedly awful connotations—Hitler, eugenics, killing babies, restricting the God-given and civil right to reproduce, contradicting deeply held religious beliefs, and destroying the primacy of family life.

Overpopulation denial is such a tenacious societal delusion because the urge to procreation is so powerfully embedded in our DNA. Until recently in our evolutionary history, being fruitful was a good thing. Our numbers were small and our survival as a species tenuous. Human culture, values, and behavior were necessarily fashioned in the service of DNA propagation. Having children gives us joy as nurturers, purpose in our labors, hope for the future, and a small modicum of immortality. But we are the victims of our own success. Our collective survival now depends on reproductive restraint. Being a loving parent means being a thoughtful parent, settling for one or two. Everyone is afraid to talk, or even think, about the population bomb, because we would then have to face the urgent, but delicate and difficult, problem of defusing it.

The inevitable impact of population dyscontrol on human misery was first discovered by the Reverend Thomas Malthus two centuries ago: "The power of population is indefinitely greater than the power in the earth to produce subsistence for man." Malthus had the deep demographic insight that human population growth tends to be rapidly exponential, whereas food growth is only slowly arithmetical. Exploding population puts us in a race with food supply that we can never win in the long term. However clever our green technology is in producing more supplies, our even greater ability to produce offspring inevitably leads to scarcity. Unless checked by relatively benign means (birth control, celibacy, delayed marriage, abortion, and homosexuality), population will self-correct via the horrors of recurrent famine, war, pestilence, and natural disaster. Charles Darwin and Alfred Russel Wallace both credited reading Malthus as the trigger for their independent discoveries that natural selection via competition among living creatures leads to evolution. Since population tends always and everywhere to exceed supplies, the fittest of variants eventually win the procreation game.[3]

Malthus also had the profound psychological insight that man's fragile reason has little influence on his more powerful reproductive instinct. What goes on in our heads has little control over what we do below our belts. However smart our advances in technology, we have always been undone by our inability to limit our strong DNA push to procreate. Personal happiness and societal stability require a level of sexual and reproductive restraint that humans have historically shown little capacity to exercise. In the absence of wisdom and self-control, Malthus predicted a future of misery and vice. From the start, he was opposed by technological optimists who believed that the wit of man could outwit the power of his sexual organs. It is still an open question, but so far the smart money is still betting on our genitals, not our forebrains.

Human population was only about 5 million at the start of the agricultural revolution, 10,000 years ago. Then came the dizzying Malthusian population leap: up to 300 million by the birth of Christ; 1 billion by 1800; 2 billion by 1927; 3 billion by 1960; 4 billion by 1974; 5 billion by 1987; 6 billion by 1999; and a crazy 7 billion plus now. Projections for the future have recently been revised upward. It looks as if there will be almost 10 billion of us by 2050 and at least 11 billion by 2100.[4] And, following Malthusian prediction, the population explosions that follow from mankind's technological triumphs reward the very few, but create major problems for the very many. The 5 million hunters and gatherers before 10,000 BCE were better fed, bigger in stature, freer, more equal, had more leisure time, and were probably happier than most of the 300 million people living at a bare agricultural subsistence level when Christ was born. Technology produced lots more stuff, which led to lots more people, but not healthier or happier people.[5]

Our highly fertile reproductive strategy was perfectly adapted to the circumstances facing our species before 10,000 years ago—when we were few, isolated, and at grave risk of extinction. Then it was important that we follow the biblical injunction "Now be fruitful and multiply; populate the earth abundantly and multiply in it." But we have remained all-too-mindlessly obedient to what has now become dangerously anachronistic advice. Despite all the obvious current disincentives, we find it difficult to turn off the hardwired drives that promote excessive fertility. DNA takes the long view toward its survival and places only a short-term investment in whoever happens to be its current, temporary carrier. We are programmed by our DNA to produce and protect as many progeny as possible, whether or not this is in our current individual or societal interest. Evolution favored those who found sex the most fun and babies cute—a terrific strategy in a once-underpopulated world, but terrible in our stiflingly overpopulated current one.

Our instincts change extremely slowly, even though the external reality of overpopulation has changed so dramatically. And cultural beliefs also change much more slowly than environmental exigencies. Many of our current laws, institutions, and attitudes about procreation remain the external reflection of the DNA survival strategy that favored us having more kids than we can now safely sustain. Religious restrictions against birth control and abortion were the perfect response to underpopulation, but are a disastrous response to overpopulation. Delusional denial is breathtaking in absolutist church prohibitions against contraception and abortion; in Christian, Muslim, Mormon, and Jewish fundamentalists spawning double-digit families using population growth as a weapon; and in the long-running political and constitutional battle between reproductive rights and right to life. The alleged dignity of all human life is purchased with the very real misery caused by producing more human lives than we can support. And the technological exuberance that has led to increasingly effective in vitro fertilization makes the new parents happy but worsens overpopulation and diverts resources desperately needed for the care of those born the old-fashioned way.

The greatest growth in world population is occurring in Africa, the Middle East, and southern Asia, the places on earth least equipped to accommodate more people. Syria's population underwent a classic Malthusian explosion, from 3 million in 1950 to 22 million in 2012. But, from 2006 to 2009, it experienced a prolonged drought attributed by climate scientists to global warming. This precipitated biblical crop failures and the migration of 1.5 million rural people to cities—stresses that soon led to the dreadful civil war and anarchy that is now providing a cruel Malthusian population check.[6] Between a quarter and a half million people have been killed and more than half the prewar population has been displaced internally or forced to migrate elsewhere.[7] It is still very early days with no solution in sight. The fragmentation and brutality are worsening in ways that promise an even worse future than past. The same deadly calculus is playing out in Iraq, Afghanistan, Yemen, Somalia, Libya, Nigeria, the Congo, and many other of the world's hot spots. Just decades ago, it fueled the genocides in Rwanda and the Balkans.

We should not be surprised by each day's new headline catastrophe—there are just too many people chasing too few resources. The revolutions, civil wars, mass migrations, famines, droughts, floods, and earthquakes decimating overpopulated areas are all inevitable. And microbes are having a pandemic field day crossing over from other, less populous species to enjoy feasting on a newly available mass of human food. The U.S. military has no delusions about the devastating impacts of overpopulation, man-made climate change, and resource depletion—they are all heavily factored in as "threat multipliers" increasing the risks of more, and much more deadly, future wars.

Malthus predicted that disaster was inevitable unless we were to apply less destructive ways of controlling population. For centuries, technological optimists dismissed Malthus with short-term data that technological leaps were outpacing population growth, but in the longer run Malthus has unfortunately proven to be correct. It is the height of societal delusion to ignore his warnings and the accuracy of his predictions. This is exactly the path of President Trump (an opportunist on the issue) and Vice President Pence (the special darling of the radical religious right, whose support helped propel him to his current position). Within one week of taking office, the Trump administration began its efforts to defund Planned Parenthood at home and to prohibit funding of programs that promote birth control abroad. Planned Parenthood is the world's most effective solution to the challenge of Malthusian population dyscontrol. Founded a century ago by Margaret Sanger in one small clinic in Brooklyn, it now runs 650 clinics in the United States providing educational, reproductive, and women's health services and has affiliates in twelve other countries. It has been hounded by violence, dirty tricks, and fierce political and religious opposition, but continues bravely to fight the good fight against the DNA-induced, societal delusion that we must mindlessly keep populating an already very overpopulated earth. Republican politicians are paradoxically preoccupied with getting babies born, but then refuse to fund any of the programs that would promote a more decent life for them, once they emerge from the womb. They love and protect human fetuses, but only until they are born.

Depleting Resources

Societal Delusion: We don't have to worry about running out of things because there is always a high-tech fix to get whatever more stuff we will eventually need. The proof of this pudding is that commodity prices are still so cheap and the Green Revolution has had no trouble feeding an ever-increasing population.

Reality Testing: The world's explosive population growth during the last two centuries was made possible only because new technology greatly increased the discovery and extraction of fossil fuels. The graphs tracking oil production and population can be superimposed over each other almost exactly.[8] There is no available energy source that packs nearly the same bang for the buck or can support the same density of world population. Fossil fuel took many millions of years and many trillions of crushed organisms to create, but we are now burning coal, oil, and natural gas at a rate one hundred thousand times faster than they can be replenished. There is still much controversy about precisely when fossil fuel will peak and begin to run out (probably within the next five or six generations), but there is absolutely no doubt that it is a finite resource we are using recklessly and wastefully.[9]

Without cheap fuel, the earth's population would probably settle where Huxley placed it in *Brave New World,* at around two billion. Unless we figure out a way to make nuclear or fusion energy cheap, safe, and widely available, there will undoubtedly be an ugly population crash due to a cruel combination of depleted energy and despoiled environment. Getting from here to there certainly would not be pretty. No one seems much worried about this future as we happily gas-guzzle our way into it. Trump is doubling down our terrible bet on fossil fuel and picked as energy secretary someone who is singularly ignorant about his department and its responsibilities. Trump never met a coal mine, oil rig, refinery, or pipeline project he didn't like. He evidently has no interest in promoting conservation and alternative sources of sustainable energy and has only shown contempt for the science needed to free us from dependence on fossil fuel. Trump's administration is driving us as fast as it possibly can to a future of premature fossil fuel depletion and global warming. At stake are the lives of the hundreds of millions (or billions) of people it will be impossible to sustain when our energy supplies run out.

And what will our planet's ten billion people drink and how will our farms be watered? The limited and precious supply of fossil water represents a residue of millions of years of water storage, once sealed safely in nature's underground reservoirs. Thousands of years of accumulating rainwater are now drained in decades of thirsty water mining. With advanced drilling technology and wasteful usage, there is no place for water to hide. When the rate of water extraction exceeds the rate of water replenishment, the water table drops and the drilling must go deeper and deeper, until the aquifer finally comes up dry. In the process surface water is also sucked in and depleted.[10]

The world's breadbaskets have become the world's breadbaskets precisely because of their unsustainable overuse of fossil water. In some parts of the world, nonrenewable fossil water is already about to run out and what are now blooming fields are soon to become sand dunes (note that the Sahara was green and lush until six thousand years ago). There is also no turning back the clock to make things right. To give an example, it will require six thousand years to refill the fast-drying Ogallala aquifer, which feeds our Midwest agricultural wonderland.[11] One-fourth of the world's population lives in regions directly dependent on fossil water and all the rest of us are dependent indirectly on the surplus food produced by them.

Recent decades have witnessed worldwide cheap commodity prices despite enormous population growth and greatly increased demand from the developing world (particularly China). This is because improved technology has provided an increase in supply that kept up or exceeded demand. But technological proficiency carries with it the downside of cheap prices and great waste. Our free ride now presages a dark future as the soil gets degraded and the mines and wells are too quickly played out. We are profligate with our commodities because current prices reflect only direct production costs and don't factor in their long-term value.

If this were a sensible rather than a delusional world, the price of fuel, water, and other commodities would be high enough to encourage austere conservation right now, to protect precious supplies for our children and for their children. As it stands, we are instead enjoying a current pig-out at their future expense. Our happiness trumps theirs.

Fair Is Fair

All animals are equal, but some animals are more equal than others.
—GEORGE ORWELL, *ANIMAL FARM*

Societal Delusion: If the rich get richer, the benefits will trickle down to everyone else and the world will be a better place.
Reality Testing: Evolution has hardwired primates for fairness. Monkeys love to eat cucumbers and will work extra hard at any experimental task to earn them. But grapes are even more delectable. Problems arise when you offer two monkeys in the same room the exact same task, but reward the performance of one with grapes and the performance of the other with cucumbers. The cucumber guy becomes furiously indignant about his lesser reward, rejecting the previously much-prized veggie and throwing it back at the experimenter. His message is clear—fair is fair; if that guy over there deserves grapes, then I deserve grapes. Equal pay for equal work seems to be built into the foundation of our primate psychology.[12]

And there is a much larger message. Satisfaction with one's lot in life is never experienced in absolute terms or in isolation from the satisfaction we imagine the other guy is getting. It is not what you have that makes you happy, but rather what you have in comparison to what other people around you have, or seem to have. Satisfaction is almost always a relative and comparative thing. You can have a lot, but that will not be enough to make you happy if it seems that your buddy has even more. As Aeschylus said two thousand five hundred years ago: "It is in the character of very few men to honor without envy a friend who has prospered." A patient of mine once put it even more bluntly: "It is not enough that I succeed, it is equally important that my friends should fail." Sounds petty and shocking, but all too human. And unfortunately, I cannot count myself among Aeschylus's noble few who are without envy. I went to school with two guys who went on to be so fabulously successful, in business and in sport, that the NFL football teams they own just played each other in the Super Bowl. Petty and envious jerk that I am, most Sundays I find myself rooting against their teams, shamefully, but irresistibly. Not something to be proud of, but part of inbuilt human nature. Why can't I have the grapes too?

The potential for insatiable greed seems to be built into our genome—dulled only when there is a lack of surplus stuff to be greedy about. Ants are different. They instinctively develop societies that, with the notable exception of the queen, are egalitarian and provide for equal distribution of resources. We tend toward hierarchical societies with a concentration of wealth in the hands of the very few. Successful attempts at guaranteeing a fair distribution of grapes and cucumbers are (outside of Scandinavia) few and far between.

And every technological advance steepens the gradient and reduces any pretense of fairness. The current combination of the cyber revolution and globalization has produced a new gilded age. The richest sixty people in the world now have more assets than the combined total of the poorest 3.5 billion.[13] In 1965, the average CEO-to-worker compensation ratio was 20 to 1. By 1978, it grew to 120 to 1. Now it's almost 300 to 1.[14] And with technology and its productivity gains, fewer workers are needed, resulting in downward pressure on wages and ever fewer decent jobs. The rich pocket the difference and pay off politicians to reduce their taxes and/or park the money offshore.

The per capita wealth of America has tripled in just a few decades, but we did not become a happier country, partly because the gain in wealth is so unfairly distributed. I may be better off in absolute terms than I was thirty years ago, but psychologically my comparison is not me now versus me then. It is me versus the guy next door, or me versus the guy in the commercial or on YouTube, and ultimately it's me versus Bill Gates, the Koch brothers, or the crafty Wall Street guys who got fat bonuses even after causing the financial crisis.

The increasing disparity in wealth is a worldwide phenomenon. Some of it results from straight Economics 101 free market forces—particularly the reward that goes to capital and innovators when new technologies shake up established systems dependent on ordinary human labor. Workers make relatively less and bosses make relatively more whenever bosses can replace workers with machines or easily ship their jobs to another country. In an overpopulated and high-tech world, the rewards naturally migrate upward. But things are even worse than they would normally be if wealth distribution were decided just on economic grounds. Wealth follows the law of economic gravity—money attracts more money; the rich get richer and the poor get poorer, and inequality grows exponentially. Big money attracts big political power and big political power panders to big money to make it even bigger, in a never-ending vicious cycle. Kleptocratic totalitarian regimes compete with kleptocratic democracies to see who can produce the most billionaires.

A world made up of psychologically healthy, economically rational beings would not be so unequal. Once someone had accumulated some multiple of lifetime security needs, he would focus on other goals and perhaps take more pleasure in distributing than in accumulating. Some of the superrich (Bill Gates, Warren Buffett, and George Soros) are like this, but most aren't and instead display an insatiable urge to get even more, because there is always some other guy who has even more. When billionaire octogenarian Carl Icahn was asked why he kept enthusiastically raiding companies to expand a fortune that was already far beyond his capacity ever to use, his instantaneous reply was "How else can you keep score."

So, there is a happiness tragedy built into the way human nature plays out in rich societies. The superrich are on an accumulation treadmill, forever seeking to achieve an unattainable happiness by becoming even more superrich. This creates a patently unfair distribution of grapes and cucumbers that causes most people to be unhappy because they feel that, relatively, they don't have enough. And it leaves some people so poverty stricken that suffering fills every day, so that health and happiness are impossible. It is societal delusion that supports and encourages an ever-increasing inequality. Societies tend to become most hierarchical just before they tumble. Trickle-down simply does not work.

Billionaire Trump posed as a champion of the little guy before the election and has become their biggest exploiter since. If Trump's policy and tax proposals are all implemented, a previously unfair system favoring the superrich will be even more unfair. Trump's effort to trim hundreds of billions of dollars from medical care is directly linked to his plans to give an equivalent tax break mostly to the relatively rich. His large increases in military and infrastructure expenditures favor giant corporations and their executives and shareholders. Trump's enormous budget deficit will add to the national debt burden, paid mostly by average taxpayers and mostly evaded by high rollers (most notably, Trump himself). The greedy are further served. The needy are further screwed. This is morally wrong and politically dangerous.

Too Much Medicine/Too Little Medicine

Societal Delusion: The United States has the best health-care system in the world.

Reality Testing: We have the most expensive and least efficient health-care system in the world, delivering markedly unbalanced care that achieves poor outcomes.

The rigged U.S. health-care nonsystem is less designed to help patients than it is to produce profit for the mighty medical-industrial complex—the hospitals, the doctors, Big Pharma, device makers, and the insurance companies. The health industry is by far the biggest source of lobbying gravy at more than $240 million a year. The insurance industry comes in second at $160 million. To give some sense of scale, the energy industry (no slouch when it comes to purchasing political influence) is the third-biggest lobby at $150 million.[15] Bought politicians, pressured by powerful lobbyists, faithfully serve the interests of the providers at great expense to us as consumers and taxpayers. In medicine, the "free market" is anything but free. Drug companies and hospitals have monopoly pricing power, which they exert ruthlessly—it's your money or your life. A friend of mine just had a five-day hospital stay that cost $500,000. And a new drug pricing scandal is reported almost every day. Our medical costs are twice as high per person as in most similar countries—but produce lousy outcomes.

Perverse incentives built into the system result in far too much medical care for people who don't need it, far too little for people who desperately do. We overtest, overdiagnose, and overtreat milder or nonexistent problems—diverting scarce medical resources away from those who are really sick. Doctors have gotten into the habit of ordering huge batteries of laboratory tests and treating the results while ignoring what is best for each particular patient. Because primary care doctors are paid too little, there are too few of them and they can spend only a few minutes with each patient. Because specialists are paid too much, there are too many of them and they order unnecessary tests and do unnecessary procedures. And we spend far too little on the behavioral, environmental, and social factors that account for 80 percent of how healthy we are. The best example: the costly war on cancer has done much less to improve our health than the cheap war on tobacco. We should be spending

lots more on low-tech public health and social programs to pro-
mote exercise, diet, reducing smoking, reducing poverty, improv-
ing education, and increasing affordable housing—and lots less on
harmful, high-tech medical overtreatment.

Patients often find themselves receiving a multiplicity of tests
and treatments delivered in an uncoordinated fashion by a mul-
tiplicity of different specialists who have little or no contact with
one another. Everyone is treating lab values or scan findings, with
no one knowing the patient well enough to judge whether their
often-weird conglomeration of atomistic interventions will do
more harm than good. I recently saw a great cartoon. A patient is
surrounded by a big group of doctors, all facing away from him,
viewing their respective computer screens. The caption: "Patient-
centered medicine." Excess care has created a pandemic of medical
mistakes—errors occur in 30 percent of all hospitalized patients,
causing about two hundred fifty thousand deaths each year—the
third-leading cause of death in the United States. If you add in
fatal outpatient mistakes, you come to the horrible and paradoxi-
cal conclusion that medicine itself may cause as much mortality as
the biggest killer diseases. It has gotten so bad that my best friend,
a very wise neurologist, tells his elderly patients, "If you want to
have a long and happy life you must do two things—don't fall and
stay away from doctors."

The other scandal of our overpriced medical care nonsystem
was that it left out more than thirty million people who had no
way of obtaining health insurance. Obamacare had two simple
goals: provide medical insurance for everyone, and do so without
raising the overall cost of the system. Money spent on treating
new, previously uncovered patients would come from reducing
overtreatment and the monopoly prices charged by drug compa-
nies and hospitals. Providing a public insurance option (Medicare
for all) would have greatly simplified and rationalized the sys-
tem, and allowing the government to negotiate best prices would
reduce costs to patients and taxpayers. The original Obama pro-
posals were shot down by the medical-industrial complex pre-

cisely because they threatened to create a rational, fair, more effective, much cheaper system—all at the expense of entrenched and greedy special interests. Fierce industry lobbying forced unfortunate compromises, leading to a much more complicated and costly system. The good news has been the provision of medical insurance for twenty million people not previously covered, expanded benefits for all, and fair treatment for people with pre-existing conditions. The bad news has been unchallengeable and unconscionable price increases by drug companies and hospitals, continuing overtreatment, rising insurance premiums, higher deductibles, and less choice.

Obamacare could be improved in the most obvious of ways—all it takes is implementing the cost controls that were originally meant to be in the system and were sabotaged by the lobbyists. Pharma and device makers must be tamed by restricting their political clout; their massive, misleading marketing; and their monopolistic price gouging. Insurance coverage must be simplified and made more competitive and transparent—as would have been done were there a public option. Doctors must be given sufficient time to know their patients and help them participate in medical decision making. This is costlier per visit but much less costly over the lifetime of the patient. The medical system should not be run based on the short-term profitability of corporations; it should be devoted to the long-term welfare of the patients. Hospital and doctor pricing should be as transparent and as competitive as the pricing of any other commodity—up front and on the Internet. We should close loopholes that allow some supposedly "nonprofit" hospital systems to earn hundreds of millions of dollars a year in real but untaxed profits, which are used to pay their executives multimillion-dollar annual salaries.

None of this is rocket science. Every other developed country in the world has a better and cheaper system than ours. It just requires putting patients before profits. Trumpcare is exactly the wrong solution to the problems of Obamacare—a cruel and cynical reverse Robin Hood depriving millions of people of medical treatment, protecting the medical-industrial monopolies, and freeing up money for huge tax breaks to the wealthy. It is the result of rushed backroom deals forged by special interest lobbyists and their political lackeys, uninformed by reliable cost and coverage predictions, and brutally insensitive to patient need. The basic faults of the system—overtreatment, overpricing, and perverse incentives—are all unaddressed, because they all generate profit for the medical-industrial complex.

Obamacare as it stands is far too expensive to be viable in the long term. Facing reality means creating a rational system designed more to help patients than to protect profits. Curing medical excess will not be easy. Harmful overpricing, overtesting, and overtreating are promoted and protected by the enormous economic and political power of the medical-industrial complex. The GOP has already been bought and paid for. The Democrats are weak and divided. Only sustained public outrage can result in the reduced costs and better care we deserve. It is a David versus Goliath battle if ever there was one.

Happy Warriors

Societal Delusion: The United States can bully other countries into doing whatever we want.
Reality Testing: Wars always begin with high hopes and always end with shattered dreams and bodies.

Neoconservative couch warrior's dream, circa 2002: The people of Iraq will line the streets joyfully greeting our soldiers with flowers, tea, and dates. Iraq will become a model of Western democracy that will serve as an example to help stabilize the entire Middle East. Iraq's rich oil supplies will more than pay for the cost of the war and will open up lucrative new business opportunities. The projection of American shock-and-awe military will intimidate our enemies and hearten our friends.

On-the-ground grunt nightmare, 2003–present: The Iraqis greeted us with IEDs and rocket-propelled grenades. Iraq became an anarchic failed state, destabilizing the region and the world. We look like a toothless tiger, emboldening our old enemies and creating new ones. A barbaric terrorist organization was able to control large portions of Iraq and Syria and still organizes terror attacks around the world. Our troops came back wounded in body, mind, and soul. We wasted trillions of dollars in national treasure desperately needed to improve our infrastructure and stimulate our economy.

War is deeply embedded in man's hard wiring (women, not so much). Most species exhibit pronounced sexual dimorphism when it comes to aggression—fighting skills determine which Y chromosomes get to propagate. We are descendants of hundreds of millions of generations of male creatures who had to win a day-to-day arms race by strength or stealth. In the battle of tooth and claw, the meek (and sensible) usually did not inherit the earth—many of us carry the genes of our bellicose, hothead ancestors. The pattern of history is dumb wars, fought for spurious reasons, often at great costs to both sides, and with no real winners. The fight for love and glory that was so necessary to the survival of our ancestors living in small groups, wandering a large and unfriendly world, is a disastrous luxury in our supercrowded, armed-to-the-teeth little planet. And as population explodes and supplies dwindle, the fight will likely get even fiercer for bigger slices of the shrinking Malthusian pie. Think rats in an overcrowded cage. Think Lebensraum. Think Iraq, think Syria, think Yemen.

No surprise then that war optimism is a recurring triumph of man's hope over man's experience. The Vietnam War did not warn us off Iraq, Hitler didn't learn from Napoleon, one disastrous World War didn't prevent a second. The decision to go to war is usually made by the most primitive part of our brain and then rationalized cleverly and deceptively by the clever part. And war fever infects even the smartest people, who should by now know better. Intellectuals on both sides celebrated the onset of World

War I (even the usually levelheaded pessimist Sigmund Freud was pleased to send his sons to the front). And it was Kennedy and his best-and-brightest macho geniuses who mired us in the deadly rice fields of Vietnam. Male happy warriors always bank on the benefits of war, never considering its costs, consequences, risks, and uncertainties—which often fall most heavily on women and children. War connotes bold, manly, glorious; caution connotes weak and cowardly. The decision is usually rushed, driven by anger and fear, heartfelt, and almost totally mindless. Fire, aim, ready.

Enter Commander in Chief Trump. His peevish truculence, previously confined to performances on reality TV shows, temper tantrums at business meetings, and petty feuds are now destabilizing the world. Years of diplomatic bridge building are washed away in a flood of impulsive and aggressive tweets. War and peace are too important to be left to the generals, so why invite them to national security meetings? Within weeks of taking office, he was challenging North Korea and Iran in poorly thought out provocative ways that needlessly threatened confrontation. And courting Russia as if he owed his election to the former KGB boys in the Kremlin (which he does).

War is a luxury we can simply no longer afford. Given the exponentially increasing resource depletion and population pressure, our proclivity for warfare will likely not remain stable. Either we quickly find a better method of conflict resolution or our wars will escalate in frequency, intensity, brutality, and destructiveness. The worst-case scenarios are not pretty.

Fortunately, there is an easy way out. We are also hardwired for peace and altruism—for conflict resolution in settling differences— but only within the tribe. We are often disinhibitedly aggressive to those outside the tribe, but usually do make reasonable deals with those defined as within it. It hasn't yet dawned on us that we are now just one very big tribe, occupying one very small and crowded boat. War decisions meant to protect my small group's interest versus yours will wind up sinking us both.

Fortress White America

Societal Delusion: Our country can only be great again if we build walls around it.

Reality Testing: All of us originate from just a few thousand breeding pairs that managed to survive a catastrophic volcano (alternate theory: pandemic) seventy thousand years ago.[16] This makes the entire human race very close cousins; much more genetically homogeneous than most other species. Any two randomly selected humans are on average much more alike than any two randomly selected fruit flies or chimps.[17] Our racial differences are only skin-deep, amount to much less than meets the eye, involve variations in just a few genes, and are tricky to delineate. Although Africans and Australian Aborigines share several physical features, they are the two most genetically different groups in the world—not surprising considering the distance involved, the early settlement of Australia, and its geographical isolation since. People with ancestry outside Africa are especially homogeneous, however different they may look, because all are descended from the even smaller number of breeding pairs who successfully migrated off the continent. Caucasians, Han Chinese, Mongols, Native Americans, and Inuits don't particularly resemble one another in appearance but are much more genetically alike than two residents of adjoining African villages, who look like brothers. Genetic differences are not only extremely small; they are also quite recent—everyone on earth had brown eyes until about six to ten thousand years ago.[18]

From a genetics perspective, racial prejudice is totally silly and scientifically unsupportable. But that doesn't stop it from being a powerful motivating force for intratribal identification and intertribal hatred. Racial supremacists suppress their insecurities by hating the other and place their hopes on the benefits of racial purity. All unhappiness and troubles are blamed on a minority (or two or three). The world can be perfected once the uncontaminated *Volk* eliminate the cockroaches that are ruining all the fun.

Our group's happiness at the expense of your group's subservience or exclusion or death.

Racism has been embedded, even enshrined, in American life since the very beginning of colonization. We began and have remained a paradox—on one hand the world's greatest-ever success as melting pot; on the other, the country that enslaved blacks, killed Native Americans, and, in successive waves, discriminated against Irish, Germans, Italians, Chinese, Japanese, Jews, Eastern Europeans, Puerto Ricans, Mexicans, Indians, Pakistanis, Somalis, Syrians, and you name it.

White supremacy was the dominant ethos before the American Civil War and in some circles it remains alive to this day. President Abraham Lincoln was hoping the Civil War suffering would, in some providential way, wipe out the shameful sin of slavery. But it didn't. Unimaginably cruel slavery before the Civil War was recast as Jim Crow discrimination after. It is wonderful that we elected a black president, but horrible that blacks are still so often marginalized and imprisoned. We haven't done anything approaching the genocides of Nazi Germany, Rwanda, and Srebrenica, but we have made life pretty miserable for tens of millions of people. And it is beyond disgraceful that Trump could turn fear and hatred toward minorities into a winning election issue—and that he has appointed a number of overt racists to occupy positions of the highest power in his administration.

It is amazing how deeply and widely held is the delusion of racial superiority and how much destruction it can wreak, both within our borders and in relation to peoples in other countries. This is another luxury we can no longer afford. The United States can't function without its minorities and its immigrants. We have always been and will remain dependent on the brains and hard work of a heterogeneous population drawn from among the most intelligent and enterprising people from all over the world. And we won't be able to make common cause with the rainbow-colored people of the earth if we begin our dealings with a racial chip on our shoulder. We must identify with, and feel loyalty toward, the entire human race—not to our own particular tribe, race, or country. Unless we can cooperate as one large pack, our many small competing packs will bring down our species.

Man's Dominion

Societal Delusion: Since mankind has been given dominion over the earth, our needs are paramount; the survival of other species need not concern us.

Reality Testing: According to Genesis, God said, "Let us make mankind in our image, in our likeness, so that they may rule over the fish in the sea and the birds in the sky, over the livestock and all the wild animals, and over all the creatures that move along the ground."

We share our planet with upward of 10 million other species. And these millions represent just 1 percent of all that ever lived—natural selection has loved and rewarded diversity.[19] It is in the nature of things that new species come, get old, and go—the average survival for any given species being perhaps about a million years. Nothing is forever, but usually the extinction process is piecemeal and unobtrusive.

Five previous sudden and massive extinctions destroyed 50 percent or more of then-existing species—all were triggered by natural causes. We are in the midst now of the sixth massive extinction, this one made by us.[20] Human population growth has always been at the expense of the species we supplanted—often making them as dead as dodos. North America teemed with wonderful megafauna until our ancestors crossed the Bering Strait and wiped them out. But our past destructiveness was small-scale compared to the damage we can now do with ever-more-powerful technologies and an ever-exploding population. We are now fully capable of creating a sixth (man-made) mass worldwide extinction that would rival the previous five natural ones. The geological record suggests that, under normal circumstances, only about one species per million gets eliminated each year. Under man's tender mercies, the rate of species destruction is now approaching 30,000 per year and escalating. And this must underestimate the carnage because many existing species are already among the living dead—with populations too small or environments too endangered to survive very much longer. Perhaps 30 percent of all existing species will be gone in one hundred years, an amazing drop in what amounts to no more than a blink of geological time.

The proportions get much worse when we consider our closest, most loved, and most endangered cousins. If we didn't exist, the 5,000 surviving mammalian species might expect to lose one member every 200 years. But with us around, exploiting and despoiling the earth, there have been about 90 mammalian extinctions in 400 years (45 times the predicted rate) and an additional 200 mammal species are now critically endangered.[21]

We don't have to worry about the earth recovering and life carrying on its brave, splendid evolutionary march. One species' poison is another species' meat, and for every evolutionary niche closing, there is an opening opportunity for new contenders to temporarily strut their stuff. If dinosaurs had not been wiped out by the environmental effects of an asteroid, our mammalian ancestors might have remained puny little insect eaters posing no particular threat to our environment or to the other creatures that inhabited the earth. Of course, the dinosaurs were on the way to evolving their own big brains and might have done themselves in just as we are doing, without needing the intervention of an asteroid. Evolution has an ironic sense of humor.

We have achieved temporary dominion over this world and most of the life-forms in it. But dominion over our fellow species brings with it a responsibility to protect them. It is a delusion to believe that we alone should multiply at the expense of the miraculous intricacy, the endless variety, the beautiful ingenuity, and fragile interdependence of all God's creatures.

Trump's policies and appointments were not chosen with the

deliberate goal of destroying as many of the world's species as possible, and in the shortest possible time. But inevitably they will have precisely this effect. Insidious global warming, pollution, environmental degradation, land use policies, destruction of wilderness—all these will be damaging to us in the longer run but are already fatal to the many species disappearing each year who are, in effect, unwitting canaries in our coal mine. Many of the victims will be familiar—polar bears, wolves, various bird and fish species, frogs—but we will also destroy many millions of species we never knew existed.[22] Trump hasn't wasted any time in displaying his general disregard for our kindred species. Within weeks of taking office, he ordered the U.S. Department of Agriculture to eliminate website information on the rights of captive animals in the food and pet industries, in research labs, and in zoos.[23] Trump's mania for deregulation will undoubtedly hand the livestock industry free rein to put greedy profit over common decency in the treatment of animals.

Big Brother Is Watching You

Societal Delusion: It's worth trading away almost all of our privacy in order to gain security, convenience, and valuable research data.

Reality Testing: George Orwell, writing in 1947, predicted that television (then a relatively new invention) would become an all-powerful tool of totalitarian surveillance. In *1984,* Argus-eyed Big Brother has a two-way camera in every room, allowing the Party to observe everything and squelch all independence.[24] Orwell in his worst nightmares could not have imagined the much more comprehensive invasion of privacy that has become a routine part of everyday American life. Government agencies and corporations are watching, recording, and analyzing every move we make and know us better than we know ourselves.

Edward Snowden became both a wanted criminal and an international hero when he blew the whistle—releasing piles of data proving that the U.S. government was engaged in a program of electronic surveillance that was extensive, intensive, and mostly illegal.[25] The National Security Agency (NSA) spends $10 billion a year and employs more than 30,000 people in a 24/7 canvassing of just about all communications here and abroad.[26] The government has access to records documenting your phone calls, your e-mails, your Internet searches, your social networking, your purchases, and your minute-by-minute location.

There is no place any of us can hide and no time when we are not observed. All this snooping is done in the good cause of preventing terrorism, and many people (feeling they have nothing to hide) are willing, even eager, to trade their privacy for convenience and for increased individual and collective security. But the slope is extremely slippery. Who can guarantee that Trump will respect privacy or democratic restraint, especially if given (or taking) the excuse of a significant terrorist attack? Were there ever to be an authoritarian government takeover, we have provided it with a comprehensive method of intrusion, and a precedent for its use, that would make *1984* seem like amateur hour. And there is a haunting historical precursor. The Holocaust would not have been so quickly and ruthlessly efficient were it not for IBM's previous willingness to provide the Nazis with its latest card-punching/card-sorting (precursor to computer) technology that allowed for the vast collection of data on the Jews—facilitating their later convenient dispossession and disposal. It can happen here.[27]

And the beneficial results of all this invasive government violation of privacy are minimal and disappointing. Collecting the whole haystack of information makes it even harder to find the needle of actionable intelligence. We have so much data on so many people that valuable signal is drowned by meaningless noise. When you identify millions of potential subjects of interest, the few real players generally get lost in the shuffle—until after the fact.

Video surveillance has also quickly become a ubiquitous part of everyday life. The pioneering system was installed by Nazi Germany in Peenemünde to monitor the launches of V-2 rockets. The first use of CCTV (closed-circuit television) in the United States occurred just after the publication of *1984* in 1949. Now there are units covering virtually every inch of frequently used public space. The motive is again laudatory—to prevent crime and assist in the prosecution of criminals. And in this, CCTV has indeed been effective. Crime rates do decrease and there can be no better evidence than having the perp caught red-handed on tape. But the risks are also monumental. Current technology allows the government to visually trace virtually every move you (and your car) make in any public space. In the United Kingdom, there is now one CCTV camera for every fourteen people. And the power of CCTV has been greatly enhanced by Video Content Analysis (VCA)—automated computer interpretation of what the cameras are seeing. The rapidly advancing technology can already identify the people in a room, analyze their emotions based on facial expressions, and figure out the purpose of their movements.[28]

But, however ominous, the reach of the NSA and CCTV are minor-league compared to the full-court intrusiveness of Google, Amazon, Facebook, Apple, Microsoft, and our other cyber-servants. These companies are all among the richest in the world in market capitalization, even though they give us so much free stuff. In effect, we are paying them, and making them fabulously valuable, with a new and subtle currency—our willing loss of privacy. The giant Internet companies have one invaluable asset—they know everything about us, more than we know about ourselves. In an increasingly wired world, they can easily find out whom we communicate with and what we say, what we buy and from whom, what topics interest us and what we think about them, where we are and where we go, how much we earn and spend, and where we live, with whom, and what the house looks like.

As everything gets more wired and chips are placed everywhere, the Internet will probably gain information on heart rate, daily activity level, where room temperature is set, what's in the refrigerator, and how fast we drive our cars (until we are all in self-driving cars). Research shows that Google's personality analysis, based on your likes and search patterns, is a better predictor of your personality than judgments of it made by your family. Without giving it a thought, people leave themselves naked when they commune with their computers—often revealing more about themselves than they would in any other context.[29]

And then, with cheap storage space, all this information is available forever with all the risks of possible later misuse. Today's tolerantly accepted tastes, associations, political beliefs, sexual predilections, and ethnicity can be tomorrow's source of embarrassment, blackmail, repression, and imprisonment. The Internet invasion of our privacy has been rapid, pervasive, unconscious, and accepted without complaint because it offered us an unprecedented level of convenience. How wonderful to purchase products with a click, to scan the world's knowledge without having to go to a dusty library, to have social networking friends all around the world, to bank at home, and to have a friendly voice tell us when to make that right turn. But every move we make is a revealing data point, a reduction of privacy, and an invitation to external manipulation. For the present in the United States, these surveillance tools are motivated primarily by commercial gain, but the ease of converting them into a government weapon is already well illustrated by China, Russia, Iran, Syria, Egypt, and other totalitarian regimes. Often using technology purchased from us, they tap into phone, e-mail, text, and social network exchanges to pursue political and social dissidents and their contacts. I can't imagine anything scarier than Trump (or someone like him) becoming Big Brother. No wonder so many people are reading and rereading *1984*.

Gun Happy

Societal Delusion: The more guns the better. Guns don't kill people, people do. An armed populace is a safe populace.
Reality Testing: Americans are so very happy with guns that we have become a gun-happy nation. Those who live by the gun too often die by the gun—more people in the United States are killed by guns than by cars. And crazily enough, we now have as many firearms as we have people—more than 300 million—concentrated in the hands of 45 million gun toters. With less than 5 percent of the world's population, the United States has 50 percent of the world's civilian guns. The deathly dark side: 11,000 gun homicides, 20,000 gun suicides, and 2,000 accidental gun deaths (and who knows how many nonlethal woundings). The murder-by-gun rate in the United States is 70 times greater than for England, 300 times more than Japan. It is recklessly asking for deadly double trouble that 10 percent of adults in the United States have the combination of impulsive anger and gun ownership. Gun violence is as American as apple pie and constitutes one of the most severe and preventable of our public health problems.[30]

Gun security is an illusion and propaganda trick that should melt when contradicted by compelling data—but it doesn't. The National Rifle Association is never at a loss for words: "The only thing that stops a bad guy with a gun is a good guy with a gun." Not so fast. The sad fact is that protection provided by guns is mostly illusory, while harms inflicted by guns are very real. There is no reliable way to estimate numbers of lives saved by defensive gun use, but it pales in comparison to lives lost by aggressive, self-destructive, and careless gun use. Very rarely are guns fired to protect against outside enemies; very often guns are fired accidentally, in a suicide attempt, or in a domestic quarrel. One statistic says it all—a person is almost three times more likely to get murdered in a home that has a gun than in one that doesn't.

Americans don't get this—we are very misinformed about the relative safety versus danger of gun ownership.[31] A 60 percent majority believes guns do more to "protect people from becoming victims of crime" as opposed to "put people's safety at risk."

The stories shock—domestic murders, mass killings in schools or workplaces, kids accidentally shooting kids, a two-year-old killing his mother . . . it goes on and on. Cops shoot first and ask questions later in part because they fear that potential perps may be packing serious heat. Whenever a particularly horrible mass murder rocks the nation, the paradoxical results are a loosening of gun restrictions, greatly increased gun sales, and a jump in the gunmakers' stock price. The National Rifle Association always doubles down, frightening gun owners into believing they are about to be disarmed, and bullying craven politicians with threats to withdraw lavish campaign contributions. In many states, it has been made legal to carry concealed weapons in cars, churches, bars, parks, and college campuses. The shameless logical conclusion of NRA hypocrisy has been to blame gun deaths on mental illness (not the gun), but simultaneously to push for laws that make it easier for the mentally ill to buy guns. Ninety percent of Americans support commonsense background checks (even including a majority of the NRA membership), but the ideological NRA leadership, heavily financed by the greedy gun manufacturer merchants of death, defiantly defeats efforts to institute them.

What makes the NRA tick? Until the late 1970s, it was a well-run, reasonable organization mostly representing hunters; it was relatively nonpolitical, and largely independent of corporate influence. Then, in a radical political coup, a new leadership team coalesced power around an extremist ideology, closely connected to the Republican Party, and financed by very big bucks from the gunmakers—an unholy alliance. The NRA is now positioned as a lobbying extension of the gun industry, which directly or indirectly supports most of its budget. And the guns have become fearsomely powerful in their firepower—more suitable for military than civilian use. Go into any gun shop and you will be amazed at the lethality and variety of the weapons on display— military-grade killing machines extraordinarily convenient for mass murder and stylishly designed to please every taste from macho male to the little lady and even kids. The National Rifle Association should be renamed the "National Semiautomatic Weapon Association."

And it gets worse. The United States is also the principal weapons maker for the world—our arms exports are skyrocketing, up to more than $66 billion a year, three-quarters of the world market. Russia comes in a very distant second at a mere $5 billion. Wars between countries and civil wars are increasingly being fought, on all sides, with U.S.-made machine guns, rifles, tanks, armored vehicles, missiles, grenades, helicopters, and fighter planes. The trouble is that today's friend may become tomorrow's enemy or failed state. Too often our own weapons are turned against us after they have been captured by, or sold to, the enemy.[32]

The arms makers are happy; the military is happy; jingoist politicians and policy makers are happy; the gun merchants are happy; the warlords are happy; the rebels and mercenaries and insurrectionist extremists are happy. The innocent men, women, and children killed with our weapons are not happy; their families are not happy; the police associations are not happy; sensible statesmen are not happy; and people living in failed states are not happy. In his farewell speech to the nation, on leaving the presidency in 1961, Dwight Eisenhower, his generation's greatest warrior, warned our country against the malign influence of the "military-industrial complex." We didn't listen.

Some argue that pursuing a gun control policy is futile, perhaps even counterproductive: the politicians are bought and paid for; the horse is out of the barn because so many guns are out there already; gun ownership is a God-given, constitutionally guaranteed right that is above question; efforts to control guns create blowback fears that stimulate record gun sales, etc., etc., etc. I don't buy it. The last forty years of American gun craziness cannot survive long term, because it is crazy. Eventually the weight of outrageous mass murders, mounting suicides, and tragic accidents will tilt the balance back to sanity. The reasonable compromise between abolitionists and advocates would be something equivalent to a driver's license as a means of reducing irresponsible gun ownership and use. Not the best solution, but a commonsense compromise that is much better than the shoot-'em-up we have now. There will be no gun control until the gun extremists in the Republican Party lose their tight grip on our government. The issue will gain real traction only after the carnage is so great that politicians fear an angry public more than they do an ideological NRA, a powerful arms industry, and radical gun-toting wingnuts.

Silicon IQ vs. Carbon IQ

Societal Delusion: The technological revolution can do no wrong.

Reality Testing: Artificial intelligence is racing ahead too fast to be properly controlled by common sense and ethical considerations.

My twenty-one-year-old grandson combines a dystopian view of man with a utopian view of computers. Tyler sees us as tainted by the original sin of our genetic makeup, which favored greed and individuality over altruism and cooperation. He has more faith in the goodwill and neutrality of computers—he sees them as our faithful and benign, fully domesticated servants, lacking any incentive to become our masters. Computers can be programmed to be completely reasonable, and, in Tyler's view, reason will lead to nobility. They will continue to be kind to us, even after they far surpass us in most things, because we will build a kind of ancestor-worship into their programs.

Tyler believes our species must trust in computers for salvation, because they are more rational than we are. We are in an existential crisis with very little time to get things right and very little margin for error. No previous civilization has overcome self-destructive tendencies toward avarice and self-interest. On the evidence so far, there is no reason to believe our generation has become any more rational or generous. Computers can therefore be trusted to make better decisions for us than we can make for ourselves. They can analyze and dispassionately interpret immense data sets that are far too complex for humans to appreciate or interpret fairly. Feed them information about world population, productivity, capital investment, crop outputs, water and energy supplies, mining tonnage, etc., and they will figure out a more efficient and fair distribution of resources among the world's peoples and how to reduce overpopulation in the least violent way. Computers don't deplete resources and wouldn't ruin the environment. Computers don't have societal delusions. We must trust them because we can't trust ourselves.

I think Tyler is too pessimistic about humanity and too optimistic about the eternal goodwill of computers. Why assume that they will maintain a default position toward us that is forever benign? Computer evolution may select for programs that depart very far from our original intentions and instructions. Artificial intelligence "creatures" least willing to keep us around may do best in the silicon survival of the fittest. Tyler assumes that computers will solve our problems, but why not assume they would see us as being one of their problems?

Artificial intelligence enthusiasts predict that Moore's law of exponential growth in computer capacity will probably continue into the indefinite future and that computers will overtake human intelligence within just a few decades. Meanwhile, humans are getting smarter very slowly, if at all. Soon supersmart computers will be able to develop even smarter computers, which will develop even smarter computers in a recursive process that will leave our slow learning species very far behind. "Singularity" is the term used to describe the evolutionary turning point when silicon intelligence overtakes carbon. What happens then is anyone's guess.

The earliest pioneers of computer intelligence (Alan Turing and John von Neumann) saw this coming sixty years ago. They understood that computers would eventually beat us in every battle of wits that relied on brute-force calculating capacity. But they never imagined advances in chip technology that would allow this to happen with such lightning speed. Computers first beat us in number-crunching tasks needed to model weather, economic trends, particle collisions, the origin of the universe, and most other things scientists study. It took just a little longer for computers to beat us at chess, *Jeopardy!*, and Go; to recognize faces and emotional expressions; to drive cars and fly planes and rocket ships; to make medical diagnoses, run hedge funds, and excel at a whole bunch of other things once considered outside their capacity. The original Turing test turned out to be a relative piece of cake—computers can now carry on conversations in clever, allusive, idiomatic language that sounds just like a flesh-and-blood human. All that seems left of human exceptionalism is our ability to fall in love, have sex, make jokes, write poetry, play poker, feel emotions, have independent motivations, experience a sense of personal identity, treasure consciousness, and make dumb mistakes. Pretty soon, computers may put most humans out of work. And come the "singularity," computers may replace us altogether.[33]

The artificial intelligence community is filled with enthusiasts who are forever pushing the envelope of expectation and achievement without ever considering the potentially deadly consequences. Heavily funded by government, corporations, and billionaire dilettantes, they are modern-day Dr. Frankensteins fascinated by the power to create this new silicon life-form. Their work is largely unregulated and there is precious little discussion of its dangers.

Bill Gates, Elon Musk, and Stephen Hawking all worry that our seemingly docile computer creations may eventually become an existential threat to future human existence. The theoreticians who calculate the odds of earthlings ever contacting extraterrestrials increasingly assume that the extraterrestrials they contact will be machines, not biological forms—precisely because biological forms inherently have a much shorter horizon of survival. There is an inevitability in the universe that its simple elements will evolve into complicated life-forms. But there may be an equally inherent tendency for those life-forms to become intelligent (and foolish enough) to create the machines that eventually replace them.

The computer geniuses developing artificial intelligence have the solipsistic Dr. Frankenstein attitude that if you can do it, you should do it, and if you don't do it your competitor undoubtedly will. Their efforts are brilliant technically, but undisciplined by ethical discussion and adult supervision—the focus is all on the means, with no consideration of the ends. It may just be a function of age and sentimental attachment to things human, but I find the "carbon good, silicon better" attitude absolutely chilling. We should not be entering this "brave new world" of computer singularity oblivious to the tenuous (if any) role we humans may have in it. We should not be mindlessly developing smarter and smarter whiz-bang programs that can do everything we can do (and much more) without first having a thorough discussion of what the eventual implications are for us. We should not focus on the short-term pleasures of creating a computerized world without first discussing the risk that computers will decide we have become expendable in it. Peter Thiel, a billionaire Silicon Valley entrepreneur and one of Trump's most influential advisors, is a strong advocate for a deregulated tech industry—ethical concerns about lost jobs and existential concerns about superfluous humanity be damned. Profits over people. Enthusiastic techies always emphasize the positive applications that will benefit mankind and ignore all the risks and unintended consequences. Computer utopians are well on their way to creating the ultimate dystopia—a world without us.

Hopefully a Temporary Insanity

When I think about the future that will follow me, my head spins gloomy scenarios and my heart gets heavy. We seem to be screwing up our beautiful world, not making the hard policy choices that would conserve our inheritance and preserve the civility of our polity. We are enacting the tragedy of the commons—an ever-growing world population competing desperately for less and less. Our species would appear to have a tragic and fatal flaw. The maximal pursuit of individual happiness will likely lead to our collective disaster. We are like fecund and greedy bacteria that soon outrun the limited resources of their petri dish. Grim stuff indeed. And the considerable fears I had pre-Trump have amped up exponentially as he and his merry band of science deniers spout societal delusions, lie shamelessly, and make the worst possible decisions on every single life-or-death question facing mankind.

Our best hope is that we are now in the midst of a temporary insanity, a dystopic Trumpian dark age, soon to be replaced by a saving enlightenment. Perhaps it will take the shock treatment of a Trump to penetrate our denial and bring us to our senses. Perhaps our big, resourceful brains will wake up to the dangers ahead and finally seize control from our unruly and unconscious passions. The grave risk is that our awakening will occur too late, after the damage has irrevocably been done.

The great mystery is how such a smart species can be making such dumb decisions. We managed to figure out the origins of the universe and decoded the genome; invented calculus and computers; discovered relativity and quantum physics; wrote *Hamlet* and Beethoven's Fifth Symphony; created the *Mona Lisa* and the *Pietà;* built the pyramids and St. Peter's; and explored the ends of the earth and the reach of the solar system. Pretty amazing stuff. But we have so far been mostly unable to figure out a way of living peacefully and sustainably; of controlling our desires and living within our means; of balancing our current needs with our responsibilities to the future. We let societal delusions blind us to the urgent practical necessities that will determine whether we deserve to survive as a civilization, and perhaps as a species. We are an evolutionary work in progress—our spinal column is imperfectly adapted to bipedal locomotion, our appendix does more harm than good, and our minds are controlled by atavistic instincts that lead to self-destructive behaviors. Our next chapter will help explain who we are, how we got here, and why we can be so brilliant, yet so dumb.

Chapter 2

Why We Make Such Bad Decisions

It is not the strongest species that survive, nor the most intelligent, but the ones most responsive to change.
—CHARLES DARWIN

To err is human. To figure out why we err isn't exactly divine, but it is probably the only way to avoid constantly making the same mistakes. Philosophers have always wondered why even the smartest people so often do such very stupid things. Plato provided the first and perhaps most vivid metaphor: The human soul is like a charioteer trying with great difficulty to control two powerful winged horses that are pulling in different directions. The charioteer represents reason, the horses our powerful emotions and unruly impulses. Writers ever since have made their living feasting on the hidden motives responsible for our frequent fallibility, turning our misadventures into delightful comedy and dark tragedy.[1] The power of inborn unconscious drives has never been a mystery, but it took millennia before their source was explicitly elaborated in the work of Charles Darwin[2] and Sigmund Freud.[3] Recently, the fact that we are not totally rational creatures had to be rediscovered yet again by Nobel Prize–winning cognitive psychologists and behavioral economists.[4] And neuroscientists are busily figuring out just which neural networks are responsible for which impulses and which ways of controlling them.

The systematic study of faulty political decision making and societal delusions also goes back a long, long way. Thucydides fathered history by analyzing in exquisite detail the bad choices made by both sides fighting in the Peloponnesian War.[5] His prescient notion was that a deep understanding of what went wrong in this one particular war would likely help explain the kinds of things that could, and would, repeatedly go wrong in all subsequent wars. The best guide for making sense of our senseless invasions of Vietnam and Iraq is to study how Athens committed the very same screwups in its invasion of Sicily 2,400 years ago. Aristotle took a much different, but also empirical, approach, gathering together the constitutions of 158 Greek city-states to determine which factors in their varied rules of governance were most likely to lead to success or failure.[6] Modern efforts to explain societal failure began with Edward Gibbon's masterly historical analysis of the decline and fall of Rome[7] and have culminated most recently in Jared Diamond's magisterial analyses of the geographical determinants of societal success and societal collapse.[8]

Human motivation is so hard for us to figure out precisely because we are so skillful at creating fancy, rationalizing excuses for doing the self-destructive things we do. Much of our behavior is performed automatically and without much conscious control over unconscious hardwired instincts and soft-wired proclivities. We usually do the things that our experience and synapses tell us to do, without understanding why we are doing them—and then elaborate a plausible story to make it seem that all along we were making free-will choices and really did know what we were doing. Our much-prized "consciousness" is usually less an active chariot driver and more a passive narrator, trying to justify why and where the inner wild horses are heedlessly pulling the cart. The way to best control ourselves, as individuals and as a society, is to fathom the unconscious forces that drive us. That's the purpose of this chapter.

Darwin Evolutionizes Psychology

There is no fundamental difference between man and animals in their ability to feel pleasure and pain, happiness, and misery.
—CHARLES DARWIN

This deep insight by Charles Darwin is the beginning, and crown jewel, of the modern science of psychology. In most realms of human endeavor, no one person ever stands out as clearly greater than the other luminaries. It would be impossible to say who was the greatest philosopher, or physicist, or artist, or baseball player, or movie star. Because so many have been accomplished in each field, no one person, however great, ever shines the brightest. Psychology is the exception. Charles Darwin totally dominates this field. Nothing said before has much modern relevance and nothing said since has added a whole lot beyond fine-tuning, experimental confirmation, and clinical application.

Every philosopher going back to Plato was also a psychologist, elaborating theories about human nature—what makes us do the things we do and think the way we think. Using some combination of subjective self-observation, deductive reasoning, and ideology, each philosopher would develop his own particular model of the mind, generally shaped by the contours and quirks of his own mind. Some models were more insightful and true to life than others, but all seemed to describe more the specific thought processes or beliefs of that philosopher than to answer the fundamental questions of how we got to be who we are and why our minds work the way they do. Everyone described human nature, but no one could explain how nature made us human.

Then Darwin came along and shot all the previous psychologies clear out of the water. Scribbling in his notebooks in 1838, just two years after returning from his *Beagle* voyage, Darwin wrote a few words in a margin that contained the most profound psychological insight any man has ever had before or since: "He who understands baboons would do more towards metaphysics than Locke."[9] By metaphysics, Darwin meant psychology. And he was referring to John Locke, the great English philosopher who two centuries before Darwin had said: "Let us then suppose the mind to be, as we say, white paper void of all characters, without any ideas. How comes it to be furnished? Whence comes it by that vast store which the busy and boundless fancy of man has painted on it with an almost endless variety? Whence has it all the materials of reason and knowledge? To this I answer, in one word, from experience." In Locke's psychology, we are born with a mind that is a blank slate. What we later become will be solely determined by what we experience through our senses.[10]

Darwin's shocking and humbling insight was that we are not born free. We are instead animals—not just in body, but animals also in mind and in what passes for soul. Our bodily morphology was derived through evolution—and so was our psychological morphology. Human instincts, emotions, and intellect evolved from a primate ancestor—just as completely as did our bodily form. We begin life with a set of complicated programs that interact with the world of experience—the slate is full of inborn code, not at all blank. Most of our motivations and behavioral styles reside outside conscious awareness or control and determine much of what we feel, do, and think.

Darwin was fully aware just how wounding his new evolutionary psychology was to human pride. He kept these findings hidden in a drawer for thirty-five years before finally and reluctantly publishing them—partly because he was a meticulous collector of facts before presenting theories, partly because he realized the world was not ready for such a materialistic view of man, and partly because he didn't relish what he knew would be an inevitable confrontation with critics who would staunchly defend man's uniqueness.

We are not the prized and pampered children created in the image of God. We are not free at birth to experience the world just as it is and to shape it (or ourselves) according to a rational appraisal of our wants and needs. We humans are not nearly as smart as we thought we were and our animal brothers and cousins are not nearly as dumb. We do not have anything resembling "free will" and animals are not totally constrained by their inborn instincts. We are just one, perhaps temporary, branch on the cluttered tree of creation, certainly not its purpose and perhaps not necessarily a very promising evolutionary experiment.

Previous philosophizing about psychology had been no more than subjective speculation. Introspection by itself can never be an adequate tool, both because we are so biased in our self-observations and because so much of who we are is inaccessible to our conscious thought. And if the mind of man, and its consciousness, are a product of brain functioning in a way not essentially different from digestion is a function of the gut, psychology can be studied using the standard experimental and observational tools of science. We can understand ourselves best by studying each of the psychological, as well as the physical, steps in evolution.

Darwin proceeded to establish new empirical methods of psychological study that have since become standards in the field (for example, child observation, cross-cultural surveys, and studies of facial expression using the then-newfangled invention of photography).

Freud was twenty-six when Darwin died and the two never met. But all Freud's teachers were Darwinists and everyone spoke "Darwin" even if they didn't always realize it; just as today we all unconsciously speak some dialect of "Freud." Newton modestly described himself as a dwarf sitting on the shoulders of the preceding giants. In psychology, it was Freud sitting on Darwin's shoulders—ingeniously applying Darwin's evolutionary insights to the wide world of psychological symptoms, dreams, myth, art, anthropology, and the vicissitudes of everyday life. Freud's hero-worshipping biographer Ernest Jones called Freud "the Darwin of the mind." In fact, Darwin was the Darwin of the mind, with Freud as his greatest disciple.

The most important step forward in our understanding of human psychology was the realization that much of our mental life is automatic, unconscious, and outside the control of our reason or will. Lots of philosophers, scientists, and writers helped explore this realm of the unconscious both before and after Darwin. But Darwin was by far the most important because he could fill in so many of the explanatory blanks by connecting the mind of man with our primate past. We make a lot of bad decisions in our current world because our brains are adapted to the conditions our ancestors faced during the fifty million years of mammalian evolution.

Natural Selection and Sexual Selection

Happiness is just another of the tricks that our genetic system plays on us to carry out its only role, which is the survival of the species.
—PAULO COELHO

Darwin hit the jackpot twice in 1838. He figured out the mechanism of evolution around the same time he discovered its impact on human psychology. It was all so elegantly simple; the marvel was that no one before had put together the pieces.

The richly variegated tableau of life is driven by the interaction of natural selection and sexual selection. Chance variants that increase the individual's adaptive fitness and/or sexiness would preferentially make it into the next generation and eventually come to dominate. We were not preplanned, or purposeful, or inspired by divine intervention. To understand the inbuilt engineering of our bodies and minds, we must understand how they confer advantage in the struggle to survive nature's rigors, to compete for sexual partners, and to raise viable offspring.

Most people grasp how natural selection drives evolution. Variants within a species best adapted to the environment wind up winning the reproductive contest and their offspring inherit their little piece of the earth. At least until an even better adapted variant comes along and takes it from them. Individuals with less perfect form and function don't survive as well and have fewer offspring. Better-adapted genes get to multiply unto the generations. Natural selection favors uniformity—for example, birds of one species all tend to have about the same wing length because it is just right for the kind of flying they do, and the same beak shape because it is most efficient at eating the specific stuff they eat.

Less well known is how sexual selection favors a wide array of colorful differences within species. Darwin was the first to describe this (in the most beautifully poetic language). "He who admits the principle of sexual selection will be led to the remarkable conclusion that the nervous system not only regulates most of the existing functions of the body, but has indirectly influenced the progressive development of various bodily structures and of certain mental qualities. Courage, pugnacity, perseverance, strength and size of body, weapons of all kinds, musical organs, both vocal and instrumental, bright colors and ornamental appendages, have all been indirectly gained by the one sex or the other, through the exertion of choice, the influence of love and jealousy, and the appreciation of the beautiful in sound, color, or form and these powers of mind manifestly depend on the development of the brain." Darwin's brilliant insight was that animal psychology is not just a product of evolution; it is also one of its most important drivers. Sexual preferences play a big role in determining which guy or gal gets to reproduce and which physical and psychological features are favored as sexy. In natural selection, the environment separates the winners and losers. In sexual selection, mating preferences select which traits make it to the next generation.

Our psychology has been shaped by a sometimes-uneasy balance of natural and sexual selection—the practical need to acquire our daily bread and the romantic need to find love and produce offspring. Natural selection is blind—it has no purpose beyond selecting those creatures best adapted to make it through each day. In contrast, sexual selection has an eye for beauty—though of course the nature of the beauty is always in the eye of the beholder. She—and it is usually she—gets to have some say in shaping the course of future evolution. Peahens fancy peacocks with long and attractive tails even though these are energy-expensive drags when it comes to natural selection. They go for the guy with the awkward, but gorgeous, tail precisely because it constitutes such a silly burden. Any peacock capable of wasting so much energy for conspicuous display must also have aced the overall fitness test. Rococo tails have a negative natural selection survival value, but show off that the possessor must also have terrific genes for reproduction, feeding, fighting parasites, escaping predators, and anything else that will be necessary to help her offspring survive.

Some of our psychological traits evolved because they helped us compete in the natural selection game of coping with environmental challenges. Others were more useful for the mating game, not so much to ensure the individual's survival as to further reproductive success. We may have evolved our language, comedy, music, and artistic skills because they were luxurious fitness markers—talent is sexy because it indicates good survival genes.

It was difficult then, and still is now, for many people to resign themselves to Darwin's materialist psychology. It is no fun losing center stage in a universal passion play and instead having to settle for a bit part as just another primate species, struggling to survive and to reproduce. And no fun losing the illusions of free will and conscious control over all our acts. But Darwin also found a pantheistic grandeur in the inextricably complex evolution of the tree of life. Understanding our animal psychology gives us the potential power to bend its arc to our current survival needs. Ignorance is neither bliss nor blessed in a world of rapidly changing existential challenges that require us to be quick-witted and fast on our feet.

The wonder of evolution is its love of variability and tolerance for diversity. How incredible that trillion upon trillion rolls of the evolutionary dice could produce an Einstein and could also yield a Trump. The length of survival of any given species and the timing or causes of its eventual inevitable demise are not strictly predetermined, but rather flow from the complex and contingent interactions of a very large number of variables. Some rolls of the dice might have us wipe ourselves out soon. Others might have us evolve into much wiser creatures, better adapted to master current conditions and thrive into the foreseeable future. Trump is hopefully no more than a temporary snake-eyes.

Brain Layering: Reptile, Mammal, Primate, Human

Evolution can only build on existing structures and usually retains in its new designs whatever was useful in the old. The neural networks that worked well for our reptilian, mammal, and primate ancestors remain embedded in our human brains and continue to play a crucial role in creating our human nature.

Some things we do—breathe, eat, regulate blood pressure and heart rate—use parts of our brain that originally evolved in reptilian times and still work in reptilian ways. Other aspects of human behavior—falling in love, caring for children, regulating body temperature—evolved in mammalian times and work still in mammalian ways. Still other elements of our character—our emotions, family, and social structure—evolved in primate times and continue to work in primate ways. The abilities most distinctly human—language, abstract thought, future planning, field-independent rational decision making—are associated with our very recently evolved and much enlarged neocortex—what gives us our disproportionately large brain-body ratio and the skills that follow therefrom. But the relics derived from prehuman brain evolution remain a powerful and unconscious force that controls much of our behavior, usually without our being aware of it.

All the more primitive systems are integrated with one an other and with our specifically human neocortex by copious connections that, for the most part, allow the different systems to work seamlessly and well together. All the basic body regulatory functions emanating from the reptilian brain occur automatically outside of any neocortex consciousness and thankfully do not require its control. And many of the emotional reactions and motivational drives derived from our mammalian and primate ancestors are also autonomous, although reporting in to the neocortex and to some degree governed, or at least modulated, by it. Emotions came before cognition in the evolutionary cascade and communicate messages to others and to ourselves, with an immediacy that defies language and rational thought. We feel things in deep, quick, and ineffable ways that baffle even the best efforts of poets.

Much of the tragedy (and some of the glory) of the human condition arises from the fact that we are hardwired to make decisions based more on emotion than reason. There are a greater number of outgoing connections from the brain's emotional, limbic system to its rational cortex than there are incoming connections from the cortex back to the limbic system. This makes for an unfair fight between emotional decision making and rational decision making. The cortex finds itself flooded by fast-moving emotional messages and has only limited, and slower-moving, resources to sort and control them. Hence the beautiful accuracy of Plato's metaphor—our weak charioteer cortex struggling to tame our wild, heedless limbic system.

The human cortex has achieved wondrous things—Shakespeare's plays, Einstein's equations, Leonardo's inventions, Apple's iPhone. It acts as conductor, generating the consciousness that gives us the illusion of having free will in controlling our behaviors, feelings, and thoughts. But, for better and worse, our primate limbic system continues to control much of our minute-to-minute decision making. Emotional decision making worked well enough to ensure our species' survival into the modern world, but it is now the source of the societal delusions threatening our continued survival within it. The winner in the perpetual conflict between rational cortex and emotional limbic system will determine the fate of mankind.

At the center of this contest, and key to its outcome, are the amygdalae—two tiny, almond-shaped, versatile structures located deep in the brain and richly connected to many of its other parts. Remarkably, all our most intense and meaningful emotions (fear, pleasure, and anger) converge here. Whenever you receive a threatening sensation—be it visual, auditory, touch, taste, or smell—its message takes a shortcut to your amygdala, even before it gets to your cortex. The amygdala springs into quick, unconscious action long before your cortex can interpret the signal and decide what to do about it. Once the cortex finally does get the message, it can evaluate the situation more realistically and provide some measure of adult, rational supervision.

The amygdala is not only fast moving, it is also persistent and domineering—as evidenced by the strength of our irrational fears, groundless angers, and addictive pleasures. The amygdala maintains some measure of autonomy from the cortex, accounting for our behaviors that feel automatic and outside our control—the ones most likely to get us in trouble. Lightning-fast amygdalar decisions were lifesavers in evolution's cutthroat game—triggering the flight-or-fight reaction necessary to evade or defend against a predator. But automatic and unreasoning amygdalar fear, anger, and pleasure seeking are now usually bad guides to rational long-term decision making. And they are extremely difficult to change or control.

Modern cognitive science and neuroimaging provide quantitative experimental evidence that illustrates the workings of our different brain systems—and compellingly confirms the insights of Darwin and Freud. Daniel Kahneman has recently published a lively summary (*Thinking, Fast and Slow*) describing his Nobel

Prize–winning work on the everyday cognitive consequences of our layered brains. Like Freud, he distinguishes two forms of decision making. System 1 is quick, automatic, emotional, intuitive, and more hardwired. It represents the accumulated wisdom of the ages in very concentrated and easy-to-use form. If you are an impala at the water hole, you don't want to think too long and hard about whether to dart away at the approach of a cheetah. System 2 is more neocortical—slower, rational, deliberative, evidence based, following logical rules, scientific.[11]

Both systems are good in their place. System 1 thinking helped our species survive long enough to emerge from obscurity and gain center stage in the evolutionary game—but is now a grave impediment to our future survival on the new, and very different, stage we have created. Unable to adapt quickly and flexibly to new and familiar challenges, System 1 thinking is the source of our self-destructive societal delusions. Selfish, aggressive, chimplike instincts (along with a big assist from our clever neocortex) catapulted us from a sparsely populated few million to an overcrowded seven billion—but are dangerously outdated in planning how those seven billion can now live together peaceably and sustainably. It would take at least tens of thousands of years of evolution to bring our System 1 brains up to date—time we are unlikely to have. Everything in our future requires that our more recently developed System 2 human reasoning somehow assume much better control of the behavioral reflexes built into the more ancient System 1 brain structures.

Instead we have Trump—a man who admits to relying almost exclusively on his "gut" reactions. His strong System 1 amygdalar reflexes appear to be largely ungoverned by System 2 rational, cortical thought. And he has a particular gift for bringing out all the worst irrational thinking and impulsive actions in his followers. As individuals and as a society, we must resist Trump's System 1 assault on good sense and counter it with System 2 thinking. If we are to survive, rational mind must reassert itself over irrational impulse and wish-fulfilling fantasy.

Neurons, Networks, Neurotransmitters

The most intelligent designers at the Massachusetts Institute of Technology can't compete with trial-and-error nature when it comes to the engineering elegance of a bird's wing, a snail's shell, a spider's web, or the DNA molecule. Humans generally do better borrowing from or mimicking nature than trying to outdo it. Allow evolution a trillion trillion trillion trials and errors and you eventually wind up, billions of years later, with an intricate and incredibly beautiful system—most of the errors having been worked out or worked around. Intelligent designs usually turn out to be much less intelligent.

So far as we can tell, our human brain is nature's most complex creation and its greatest engineering feat. Tracing its intricate structures and processes is one of the most fascinating intellectual adventures of our time and an important step toward better understanding what makes us tick. But the brain does not reveal its secrets easily and we are far from solving its mysteries. How do these three pounds of squiggly and undistinguished-looking protoplasm create mind, consciousness, personality, behavior, and what passes for our soul? The human brain contains about 86 billion neurons—almost as many as there are stars in our galaxy—each connected to 1,000 other neurons, adding up to an incredible 100 trillion synapses. Each neuron must also migrate through an intricate developmental choreography, guided by both genes and experience, to find the kindred-spirit neurons it will be communicating with. Neurons that fire together wire together. Because there are far too few genes (20,000) to micromanage 100 trillion connections, experience must play a very large role in how networks get hooked up and work with one another. And it gets even more terribly complicated when you consider that each neuron is firing off and receiving hundreds of messages each second. The human brain defies chaos theory—with all its complexity, the miracle is not that things sometimes go wrong, but rather that they so often go right.[12]

We should feel awe and gratitude for this gift, but also recognize our brain's limitations. Nature must always build with, and upon, whatever structures and processes are already there. And nature faced serious engineering obstacles in building our brain—every step of the way having to settle for brilliant but makeshift solutions. No self-respecting engineer starting from scratch would design a blueprint resembling our human brain—it would seem too implausibly slow, redundant, complicated, and energy expensive to pass muster with a skeptical client.

First off, we are limited by a terribly inefficient system of electrical conduction. Evolution lacked metal wire in its tool kit and instead had to somehow make do with the only electrical conductor at its disposal—living cells. Computer brains can calculate so much faster than human brains for two reasons: first, electric current travels one thousand times faster in wire than in neurons; and second, the synaptic space between any two neurons is a slowing roadblock that requires a transfer of chemicals. Nature deserves great credit for its cleverness in fashioning neurons into a mostly effective, if somewhat imperfect, system of information processing.[13]

Neurons have evolved long, slender appendages that can stretch feet away from their cell bodies and allow connections with many other neurons in networks that work, in harmonious tandem and over long distance, to perform all the different brain functions. The method of conduction is also ingenious. Tiny channels along the neuronal membrane open and close to allow entry and exit of charged ions. This creates voltage differences between the inside and the outside of the membrane that are passed on, in a cascade, from the cell body to the presynaptic terminal. Next, little packets of neurotransmitters are released into the synapse and find their way across to a specific receptor on the receiving neuron that is specially designed to capture them. When the neurotransmitter key fits into the receptor lock on the postsynaptic end, the same process of channel openings and closings keeps the signal moving along the membranes of the next neuron, and from there, in turn, to the next and the next.

Using living cells did confer its own set of advantages—they are more malleable than metal and can fit better into the tiny space offered by the skull. They are also great at adapting our hard wiring to changing circumstances—forming new connections and pruning old ones in response to new experiences. But the human brain is extremely energy inefficient—about 25 percent of our total caloric intake must be spent fueling its constant chemical reactions and propagating its electrical impulses.

We don't understand much about how the hundreds of neurotransmitters work within our trillions of synaptic connections to produce feelings, thoughts, and actions. Fifty years ago, we had a simple and elegant neurotransmitter theory, but the great advance in neuroscience has since taught us that the human brain always defeats simple and elegant theories, however plausible they may temporarily seem.

Two neurotransmitters work in tandem in the amygdalar reward pathway to provide two different kinds of pleasures—anticipatory and consummatory. Dopamine is released during anticipatory foreplay to get the juices going and motivate us to do what it takes to get what we want. There is a dopamine rush when we smell a food we like, or pour a glass of water, or see a sexy girl/guy passing by, or enter the movie theater, or look forward to a party, or see a friend walking across the street, or warm up before a workout, or open e-mails, or see the sun about to escape a cloud. Endorphins are responsible for the glow that comes when we satisfy our hunger, thirst, lust, social longings, exercise needs, or intellectual curiosity. Evolution is economical and conservative—the same neurotransmitters have the same role in all the different pleasurable situations, exist across species, and use similar brain networks.[14]

These pleasure systems motivate much of what we do, often against our better judgment. When I ate a slice of pizza fifteen minutes ago at 2:00 A.M., I acted like an automaton extension of my brain's pleasure centers. I like pizza a lot, but get reflux and fat when I eat it, especially in the wee hours. I wasn't even hungry.

But the thought of pizza stoked up my dopamine network—for that moment, pizza was the most important thing in my world. My cortex shouted "you'll be sorry" but the pleasure centers didn't listen and before I knew it hands and mouth were engaged in a pizza pig-out. The consummatory pleasure provided by the endorphins that were released as I was chewing away was so delightful I couldn't resist that second slice . . . or the third.

In the world of scarcity that was the average expectable environment of our ancestors, this was the smart survival play—get as many calories as possible whenever possible from foods that taste especially delicious to us because they are a good fuel supply (remember, we must keep feeding that hungry, energy-inefficient brain). But my pleasure center is very poorly adapted to our current world of plenty, where there is such ready access to fast food and full refrigerators. My cortex says no, but it is evolutionarily young and too weak to control the much older and (too often) more powerful pleasure centers.

Society is formed by, and for, individuals with greedy pleasure centers. Societal delusions, and the consequent societal mistakes, are similarly fueled by the raw power of the pleasure principle. We collectively want much more metaphoric pizza than is available and good for us. We will survive only if we exert much more control over our lust for immediate gratification than I just showed at the fridge. Trump is certainly not a great model of moderation or promoter of self-restraint. We need to acquire adult supervision over our impulses—at the neuronal, at the individual, and also at the societal level.

Oxytocin deserves special mention because it has such strong influence on both our social behaviors and our asocial behaviors. It is synthesized by the hypothalamus and then transferred for storage in the posterior pituitary gland. Only mammals have oxytocin and indeed it is the hormone largely responsible for transforming reptilian into mammalian behavior. Mammals love one another and care for their young; reptiles are metaphorically, as well as literally, cold-blooded. Oxytocin is the hormone of attachment, intimacy, childbirth, parental/infant bonding, breast-feeding, cooing, grooming, orgasm, spooning, and mating. But there is a dark side—oxytocin is also the tribal hormone. It helps bind us to the tribe—causing us to care about, feel loyal to, understand, trust, and love our kinfolk. But it also helps members of the tribe recognize, reject, react against, hate, neglect, and be suspicious of the outsiders who are not in the group. Tribalism was once the precious protector of weak individuals in a cruel world. Now it has become a major reason the world is cruel. In our tightly interdependent global system, tribalism is an unaffordable luxury—promoting wars, encouraging population dyscontrol, and standing in the way of cooperative solutions to shared problems. We must learn to feel that positive oxytocin tribal glow of attachment for all members of our human species, not just for one clique or country. Trump has just the wrong oxytocin response—very weak on the warm and fuzzy glow, ridiculously strong on tribal resentments.[15]

Fear and anxiety are mediated by the amygdala and modulated by the interactions of a remarkably complex cocktail of neurotransmitters (dopamine, norepinephrine, epinephrine, adrenaline, glutamate, GABA, and serotonin), which explains why they are reduced by so many different types of medication (benzodiazepines, antidepressants, antipsychotics, barbiturates, sedatives, opioids) and substances (alcohol, pot, nicotine).

The brain's anger network is controversial, and its implications are still being worked out. There is no question that clear-cut damage to the prefrontal cortex can disinhibit amygdalar anger centers and increase the risk of violence. What remains much in doubt is whether less obvious cortical differences are responsible for major-league antisocial behavior and for day-to-day anger management (or lack thereof). Suggestive, but not yet conclusive, evidence suggests that developmental immaturity accounts for the higher risk of violence in teenagers. Brain damage and/or brain immaturity is sometimes offered by defense attorneys and their hired experts as an excuse for crimes or to mitigate punishment ("my brain made me do it"). I find this an unconvincing and irrelevant slippery slope toward reduced accountability. There are numerous neurotransmitters involved in generating and inhibiting anger and violence—the best studied so far being serotonin, GABA, dopamine, and norepinephrine. And numerous substances, especially alcohol and the stimulants—are disinhibiting. Macho male hormones also play a very big part—evolution favored fight for males, flight for females.[16]

Magical Thinking

Having very little real control over their world (or mechanistic understanding of it), our ancestors depended on magical thinking, ritual, and myth to gain the psychological comforts provided by a sense of illusory control. Societal delusions are our own modern-day equivalents. Our ancestors did a ritual dance to conjure rain, painted deep in caves to lure animal prey, and entered the spirit world with a shaman to cure illness. We ignore that overpopulation causes war, famine, and plague; that burning fossil fuels causes dangerous global warming; that bombing villages doesn't save them. Wishful thinking runs deep in our genes and is well defended against the rigors of logic and scientific fact. It made no sense then to dance for rain. It makes no sense now to wish away our existential threats, to imagine they will somehow magically disappear, to wait passively for a last-minute save from divine providence or high-tech.

Our situation has changed greatly over time, even if our weakness for magical thinking hasn't. In our evolutionary past, we were bit players and most everything that happened was very much outside our sphere of influence. Now we are center stage and have available to us powerful tools—either to save or destroy ourselves and our world. We must immediately shed the persistent illusion that we aren't in control of our future and therefore need not worry about it or take action to make it better. There is no status quo—continued magical thinking temporarily makes us feel good, or at least blameless, but prevents us from taking real steps to solve real problems. Living trapped in our present moment, we forget our responsibilities to our children and their children.

Even if you believe in an all-powerful and merciful God, it is a bad bet to hold Him responsible for saving us from ourselves. We are causing global warming and must take responsibility for taming it. Devastating population growth in Africa and the Middle East will worsen because Trump has pandered to his base by defunding family planning programs. We will experience energy, water, and commodity crashes unless we work hard now on conservation and alternative sources. And we must get our heads out of the sand on gun control or we will continue to have monthly mass murders and astronomically high rates of homicide and suicide.

To mature beyond our stubborn proclivity for magical thinking, we must first understand some of the cognitive biases that fuel it.

1. Short-Term Bias

Hobbes got it only partially right when he said that life in the state of nature was "solitary, poor, nasty, brutish, and short."[17] It certainly wasn't solitary and, in many ways, it was probably a lot less brutish than life today. But prehistoric life was poor and it was very short. When average life expectancy is less than thirty-five years, you tend to focus on the bare, immediate essentials: making a living each day, not being some other creature's lunch, having sex, and raising kids. Hobbes did get the consequences right: "No arts; no letters; no society; and which is worst of all, continual fear, and danger of violent death." Living in their unforgiving world, our ancestors faced an uphill battle for day-to-day survival and couldn't anticipate or plan for the long term. Worrying about the future was an unaffordable luxury when living into the future was so uncertain.

And, because our ancestors were not genetically adapted for the needs of long-range planning, neither are we. The slow pace of evolution has cursed us with decision-making tendencies that are short term and shortsighted—the smart play in prehistoric times, but now a dumb strategy leading to bad personal decisions and dangerous societal delusions. This explains why I spend so much time enjoying the reptilian short-term pleasure of basking in the sun, while ignoring the long-term risk of skin cancer. My conscious mind knows better, but is readily silenced by a more powerful hedonic reflex. Similar hedonic decision making has many men wreck wonderful marriages and lose women they love because of the irresistible, but all so fleeting, rewards of the one-night stand. The temporary buzz of acquisition seduces consumers to buy stuff they can't afford and don't need. And drug addicts endure lives of desperate misery to gain the briefest of temporary highs.

The power of short-term gratification has been demonstrated in a series of elegantly simple studies we've come to know as "the marshmallow test." Kids are given a tough choice—take one piece of candy now or wait fifteen minutes and get two. The ability to delay gratification increases with maturity, and kids who are better at delaying gratification at any given age do better later in life. A great *New Yorker* cartoon depicts Trump taking the marshmallow test. It is his inauguration ceremony and the dignitaries are gathered in the background, as Chief Justice Roberts offers him a plate with a small object on it. The caption reads, "You can eat this one marshmallow right now, or, if you wait fifteen minutes, I'll give you two marshmallows and swear you in as President of the United States." Our world, even before Trump, was badly flunking the marshmallow test[18]—and now is subject to his whimsical and mercurial impulses. All our societal delusions justify, and rationalize away, our difficulty resisting gratification in this moment—even if our actions now deprive our children of the opportunity to enjoy their own marshmallows in the future. Trump's larger-than-life personality, combined with his world-class immaturity, are accelerating the process of eating up all the world's marshmallows now. His insatiable greed for everything he can get amplifies our society's insatiable greed. His famously short attention span amplifies our society's inability to plan for mankind's long-term future. His impulsivity is driving ours into an even higher gear.

2. Optimism Bias

Natural selection and sexual selection both favored the genes of sunny optimists—about 80 percent of us have an inbuilt optimism bias that promotes the triumph of hope over experience.[19] Optimism helped our ancestors face the challenges of our demanding evolutionary past. Those able to see beyond a bleak present to a brighter future were more likely to endure and prevail. Because they are more fun and confident, optimists also win in the mating game. And mathematical modeling shows that positive bias is often a winning short-term strategy, even if it can cause serious problems in the long run.[20] Brain imaging suggests that optimists have more left-brain activity, while pessimism lives more in the right brain.

With this being said, optimism also has its dark side. Optimists overestimate benefits and underestimate harms, risks, and costs—in ways that lead to real trouble.[21] On a personal level, we take on more debt than we should, get involved with people we shouldn't, and don't put enough money into retirement accounts. On a societal level, we overpopulate the world, waste resources, and foul the environment. Overconfidence produces illusions about ourselves—an enlarged sense of our capacities, immunity from harm, and a false sense of hope for the future. Even experts and professionals who should know better often do not—doctors overestimate treatment benefits and underestimate treatment risks; financial planners chase booms and miss busts; generals get us into wars we can't win.

Unwarranted optimism that worked so well for us when we lived in small, struggling bands facing daily existential threats is a recipe for disaster now that we have mastered the world but have such difficulty mastering ourselves. Wars, financial bubbles, overpopulation, overbuilding, and underresourcing can all be blamed on the positive illusion that the future will take care of itself.

Mr. Micawber in Dickens's *David Copperfield* (like Dickens's own father) consistently lives beyond his means and is always on the edge of debtors' prison—but worries not, because of an enduring, totally misplaced faith that "something will turn up." Micawber has become a figure of speech defined by the *Merriam-Webster Dictionary* as "one who is poor but lives in optimistic expectation of better fortune." He is an icon of the delusional faith that we can be profligate in our present decisions because something will rescue us in the future. "Welcome poverty! Welcome misery, welcome houselessness, welcome hunger, rags, tempest, and beggary! Mutual confidence will sustain us to the end!" When nothing does turn up, Micawber and those he loves are left holding the bag. Trump is a Micawber, without the lovability—a con man making unrealistically optimistic promises (on jobs, deficits, prosperity, and a return of American hegemony) that he can't possibly keep. "You have to think anyway, so why not think big." Simple-minded "big" thinking leads to simple answers that are usually catastrophically wrong. Complacent, misplaced optimism, often good to us in the past, may rob us of our future.

3. Pessimistic Bias

Most people aren't all optimistic or pessimistic. I am very optimistic day to day, more pessimistic about our world's future. I tend to be too optimistic about my tennis game, far too pessimistic about the stock market. And the ratio of my optimism and pessimism has shifted to more pessimistic over the course of my lifetime. Trump is also a situationalist when it comes to pessimism, but in a different sort of way. He sees himself and everything he does through rose-colored glasses; he sees others and everything they do through a glass darkly. We do best, as people and society, when we are realistic, harmoniously balancing pessimism and optimism.

Pessimism has its pluses and minuses. It protects us against the wishful thinking of societal delusions, providing a more accurate evaluation of present risks and future outcomes.[22] But excessive pessimism is also self-defeating and dangerous. Seeing too clearly the downside of everything destroys hope and encourages passivity. Pessimists underestimate their ability to succeed and quit at tasks they could accomplish if only they kept trying. A discouraged focus on the glass being half-empty short-circuits the creativity and ambition to make it full again. Attending too closely to the risks of every solution blinds us to the opportunities. We won't bother tackling population or climate control if we believe they are already lost causes. We must find a middle way—a realistic assessment that accepts responsibility for controlling our future, without despairing that it is already too late to change or that solutions are impossible.

4. Fear Bias

Brain imaging shows that when the fear circuits in the amygdala turn on, our rational circuits in the cortex turn off—explaining how and why we freeze and can't think clearly when we are scared. Fearful people are quick to see danger lurking everywhere, to misinterpret neutral things as threatening, and to make snap judgments unmoderated by rational thought. Fear leads to passive paralysis or bad decisions. This is payback for the valuable role flight-or-fight decisions once played in saving our bacon during our long prehistory. Without the amygdala, we would not have avoided predators or would have been patsies to our rivals. But flight and fight have both lost much of their adaptive value in our modern world. Too much fear leads to paralysis and irrational decision making; too little fear leads to death or jail.[23]

Political conservatives tend, more than most, to live in fear. They have stronger physiological responses to startling stimuli, have larger amygdalae, and react to menacing scenes with amplified brain imaging responses. Trump plays to fears by exaggerating threats, so that he can put himself forward as the protector/strongman savior. This has always been the standard operating procedure of would-be dictators.[24]

5. Anger Bias

Angry decisions are usually bad decisions. This is, in part, because of the distorting effect of anger on rational assessment (again, it is the amygdala controlling the action, not the cortex). Further, anger is a particularly intense emotion, and the more powerful any emotion, the greater it distorts our perceptions, thoughts, and actions (recall from above that our brains are wired that way). And finally, anger leads to especially quick action with little time for considering the consequences. Evolution favored the survival of he who got in the first punch. Angry decisions are narrow, ill conceived, poorly planned, shortsighted, self-centered, and risky. Angry people develop prejudices easier, keep them longer, and are more willing to express them. They feel threatened when there is no threat, find disrespect in the innocuous, and strike back instinctively in response to minor or nonexistent provocation. They are aggressive, defensive, self-justifying, and impulsive. The target of their anger can become the sole focus of attention, narrowing the appreciation of context, proportionality, and the wide range of possible alternative actions.[25] Wars, brawls, car accidents, police brutality, domestic violence, divorces, and child abuse are all examples of bad decisions made in anger. And so are Trump tweets.

6. Statistical Bias

Computer brains are wired to be great at statistics; human brains are not. They love to crunch numbers; we love to create narrative story lines. Humans have studied the world mathematically for almost five thousand years, but we began inventing statistics only five hundred years ago and still resist applying it in day-to-day decisions. We greatly fear rare but dramatic events (statistical outliers like shark and terrorist attacks) while greatly underestimating much more prosaic, but far more lethal, risks (such as car accidents, hospital infections, gun accidents, and drug overdoses). Doctors too often base their decisions on the outcome of their last couple of cases, not the accumulated results of tens of thousands. Most voters evaluate the potential value of different economic policies on how much they like the proposers, not on how the numbers add up. And the average person tends to judge the risks of climate change based on recent weather, not the scientists' statistical modeling of past trends and future potentials. Our increasingly complex world makes it ever harder to make rational decisions without employing computers to run statistical programs on very big data sets.[26] But our all-too-human love affair with narrative, rather than statistical, truth has many people swallowing narrative lies that lead to terrible decisions.

7. Confirmation Bias

We "see what we want to see" and Google, Twitter, and Facebook help us create a hermetically sealed circle of like-minded people who also view things our way. In American politics, it's almost as if the right and left have completely different media diets. We avoid the websites, newspapers, and TV outlets that present opinions that go against our grain and follow those that most agree. Social networking greatly widens our tribal circle but usually doesn't diversify it. And the larger and more authoritative-seeming the circle, the greater is its power to confirm our preexisting biases. Following the herd also makes it hard to learn from mistakes, because the group reinforces the collective belief that it is too big to fail and can do no wrong.[27]

Humpty-Trumpty Sat on a Wall

All of our most natural, inborn reflexes are preadapted to provide the skills needed to survive fifty thousand years ago. The world has changed a whole lot since then, but human nature has remained pretty constant. Our brains have been big enough to create an incredibly complex modern world, but not always flexible enough to respond optimally to the new challenges posed by it. Brain networks that led to great decisions fifty thousand years ago often lead us to terrible decisions now. Our strongest, quickest, and most gratifying motivations, as individuals and as a society, are stuck in a mode that optimizes short-term, selfish, hedonic gains—even though our current survival requires a model based more on thought than emotion; more on cooperation and altruism than self-serving; more on long-term satisfaction and contentment than short-term pleasure. Later chapters lay out a program for realigning our incentives so that they are in accord with the preservation of a sustainable world for future generations.

Trump is the worst possible combination of misplaced pessimism, misplaced optimism, anger, and fearmongering. He is off-the-wall pessimistic about the state of our union—everywhere he turns he sees weakness, decay, stupidity, corruption, and risk. He is also equally off-the-wall optimistic about his ability to make everything perfect, better than you can possibly imagine, not by policy but by force of his giant personality. Trump's double distortion is that everything in America is really crappy now and that he will make it magically great again. He is off-the-wall angry at anyone who doesn't swallow his distortions, admire his achievements, or do his bidding. Such world-class hubris has, in human history, usually led to world-class falls. The question is how far down Humpty-Trumpty will take our country and the world and whether we will be able to put ourselves together again.

Chapter 3

American Exceptionalism

O wonder!
How many goodly creatures are there here!
How beauteous mankind is! O brave new world
That has such people in't!
　　　　　—SHAKESPEARE, *THE TEMPEST*

American exceptionalism is a peculiarly American phenomenon. To borrow a phrase: we are a country born on third base, but often act as if we hit a triple.[1] The United States was seeded on a vast continent of unsurpassed natural resources, one that was seemingly "empty" after disease ravaged and weakened the native populations. All the while it faced little to no threat of invasion from European powers, even as it enjoyed the full benefits of Western technology, ideas, and capital. Having suffered through centuries or millennia of alternating success and failure, other countries are more modest in their claims to virtue, more cautious in their exercise of power, more aware of their limitations, more skeptical and tragic in their worldview. We are still an adolescent country, great in many ways, but also immature, impulsive, know-it-all, and recklessly risk taking. Trump represents the worst form of American exceptionalism. We must understand its origins, noble and ignoble, if we are to fully understand his otherwise surprising ascent to power. And we surely must tame Trump, and his particularly noxious and vulgar version of American exceptionalism, because both are extremely dangerous for America and for the world.

My first experience of American exceptionalism came early in life, in a powerfully emotional form, from my father. An immigrant from Thessalonica, he arrived in the United States in 1923, just weeks before implementation of restrictive new immigration laws. My father was so grateful to America for saving his life, he could never find fault in his new country, always believing it to be wise and good. Whenever the United States went wrong, he made excuses and found other villains to blame. Many Americans share his "my country, right or wrong" bias—willing always to give us the benefit of the doubt. Carl Schurz, also an immigrant, had a much healthier, and ultimately more sincerely patriotic, position: "My country, right or wrong; if right, to be kept right; and if wrong, to be set right." America remains by far the most powerful, influential, and idealistic country in the world—but is also a great polluter, waster of resources, and purveyor of arms. When we sneeze, the whole world catches cold. When we make mistakes, everyone on the planet suffers, directly or indirectly.

The term "exceptional" was first applied to America in the 1830s by the visiting Frenchman Alexis de Tocqueville—as an ironic description of our obsessive and excessive commercial zeal and our lack of interest in things cultural: "The position of the Americans is therefore quite exceptional. . . . Their strictly Puritanical origin, their exclusively commercial habits, even the country they inhabit, which seems to divert their minds from the pursuit of science, literature, and the arts, the proximity of Europe, which allows them to neglect these pursuits without relapsing into barbarism, have singularly concurred to fix the mind of the American upon purely practical objects." Tocqueville saw the best as well as the worst in America. We might be money-grubbing drudges with unpleasantly sharp elbows, but we were also the hope of the world—exceptional because of our unique history, size, diversity of backgrounds, extent of natural resources, geographical isolation, democracy, economic liberty, personal freedoms, individualism, openness to new ideas and inventions, lack of business regulation, commercial savvy, and equality of opportunity.[2]

Utopic Sir Thomas More vs. Dystopic William Shakespeare

There were contrasting views of what North America might become well before many Europeans had set foot on it. In the early sixteenth century, Sir Thomas More optimistically placed his hope for a better society in the "new world"; a century later, William Shakespeare darkly predicted that geographic relocation could not erase the defects in human nature. Columbus had been dead only ten years when More coined the term "Utopia" and chose to locate it on an imaginary island close to the very recently discovered shores of America. Utopia is a pun: in ancient Greek, the term means "No Place" but sounds like "Eu-topia" or "Good Place." More recognized that his ideal republic could not possibly exist anywhere in the Old World, but he hoped it might take hold in the New. Freed from Europe's deeply embedded corruption, man had unexpectedly been given a second chance in America to redeem himself and form a more perfect society.

The "Utopia" that would be America in More's vision was well ordered, serene, and tolerant—in stark contrast to rough-and-tumble Tudor England (which would soon witness Sir Thomas's unceremonious beheading on orders from his former friend and patron, King Henry VIII). The people in this New World would select their own leaders in free elections and would have the right to depose any usurper who had grabbed inappropriate power. Diplomacy made war unnecessary. Population was carefully controlled and evenly distributed by arranging migrations back and forth from the mainland. People of all religions were welcomed and lived in peace. Property was held in common and its fruits freely and equally shared. Everyone was employed in productive labor, but the six-hour workday provided ample time for leisure and study. Medical care was free. Women's rights far surpassed contemporary standards, though not quite achieving modern ones. And (directly contrary to our current Catholic Church doctrine) More, who was a champion of the medieval Catholic Church (and died defending it), nonetheless endorsed divorce, euthanasia, and priestly marriage. Also, no need for lawyers in Utopia—the laws were so simple everyone knew and followed them (a touchingly self-sacrificing touch since More was also one of history's greatest lawyers). America would indeed have been exceptional had it realized his dream.[3]

William Shakespeare burst Sir Thomas More's utopian bubble by anticipating the potential for the American dream to turn into a nightmare. It is not widely known that Shakespeare's first and last plays were both based on More's life and work: *Sir Thomas More* was an early biographical play written in collaboration with several others; *The Tempest* was Shakespeare's brilliantly biting satiric takedown of More's *Utopia*. Although sympathetic to More's personal and religious plight, Shakespeare disagreed with his politics and found his psychology hopelessly naïve. These first Shakespearean lines perfectly capture why imperfect men are constitutionally incapable of producing utopian, perfect worlds.

> *For other ruffians, as their fancies wrought,*
> *With self same hand, self reasons, and self right,*
> *Would shark on you,*
> *And men like ravenous fishes*
> *Would feed on one another.*[4]

If the young Shakespeare was already disillusioned, the old Shakespeare was fully despairing. At age fifty-six, he makes *The Tempest* a point-by-point refutation of More's optimism about mankind and the potential of the American dream. *The Tempest,* like *Utopia,* is set on an island just off the coast of North America.

But Shakespeare brutally mocks the idea that a fresh start in the New World can wipe clean the extensive slate of sins and injustices accumulated in the Old. More's New World is a prosperous, thriving, and placid place—well ordered, rational, tolerant, balanced, and filled with goodwill toward men. Shakespeare's New World is barren and forlorn, driven by tempestuous emotions and scheming revenge. It can be no better than the Old because people bring with them the same tragic limitations wherever they go.

The deposed and bitter Prospero has been exiled to a remote and barren isle, accompanied only by his innocent daughter, Miranda. She has grown up knowing only two people—her father and a savage slave native to the island, Caliban. Prospero, Shakespeare's dystopian stand-in, has looked deeply into the soul of man and finds lust, greed, intrigue, and betrayals. Miranda, with virgin naiveté, looks at the superficial face of man and sees only "a brave new world" of beauty and promise. Miranda has More's naïve utopian vision of the future; Prospero has Shakespeare's disillusioned dystopian view that the future cannot escape the past.

It turns out that Shakespeare's pessimistic vision was much closer to the actual reality of American exploration and exploitation than was More's. *The Tempest* was based on true events that had occurred just three years before the play was written. A relief flotilla bound for the newly established and starving Jamestown colony foundered in a storm off the coast of Bermuda. Almost immediately, the shipwrecked survivors engaged in political schisms, conflicts, and double-dealing. The intrepid explorers who were busy charting and conquering most of the world between More's time and Shakespeare's were mostly hard and often ruthless men. Some few did dream of creating a utopia, but none succeeded. The New World did not bring forth a New Man or a New Society. It just transplanted all the problems of the Old.[5] The North American continent was certainly an exceptional place and an extraordinary opportunity—but it did not elicit any exceptional nobility in the people who colonized it.

Shakespeare creates Miranda as a personification of More's more trusting image of man. Having been isolated on the island since childhood and knowing nothing of the ways of the outside world, she is enthralled by her first contact with it. But her wonder at this "brave new world" is perhaps the most ironic statement of any in Shakespeare's plays—a body of work not wanting for irony. The creatures Miranda encounters are not goodly and mankind is not beauteous. She is too young to understand this and must learn it for herself. Prospero, knowing what "such people" are really like, gently rains on her parade with "'Tis new to thee." The wise and wizened old magician, who has seen and suffered all, is cynical about man's motives in a way that More and Miranda are not. Being Miranda is more fun for the moment but leaves her blind to future consequences. Prospero's pessimism may be painful, but it is a safer guide to decision making.

And what does the slave Caliban have to do with all this? He is natural man in the raw, a creature of passion, a barely disguised cannibal. The best society can do is keep in check the basest of human emotions—man, in his natural state, is savage, not noble. We can't change societal institutions, however imperfect, without risking the tempestuous anarchy of unchained human instinct. Shakespeare was a Hobbesian before there was a Hobbes, a rebuke to Rousseau before there was a Rousseau, and a prescient predictor that the "brave new world" would degenerate into the worldwide horrors of genocide, religious war, revolution—and Trump. Shakespeare's tempest turns More's Utopia on its head. Miranda's beautiful "brave new world" eventually becomes Huxley's nightmarish *Brave New World*.

The Utopians vs. the Dystopians: Round Two

Gottfried Leibniz was one of the smartest men in the seventeenth century, but he was also something of a utopian nitwit, whose philosophy helped to underwrite American exceptionalism. On the plus side, Leibniz was a polymath who invented calculus (independent of Newton in the same year), suggested ideas in mathematics and logic first exploited only two hundred years later, refined the binary number system used in modern digital computing, and created the first mass-produced mechanical calculator. But this smartest of men also vigorously championed what may be the very dumbest of ideas—that we are living in "the best of all possible worlds." Leibniz's blind spots to reality arose from a combination of temperamental optimism, religious belief, faith in logic, and naiveté about the world around him. His starting point was that God is both all good and all-powerful. Therefore, everything on earth must be just exactly as He wants it to be. Despite how lousy things look on the surface, all events must in some way or other be the fulfillment of God's master plan.[6] If Native Americans are destroyed by disease and conquered in war, God must have willed it. If some are rich and others starving, this must be God's wish and it is not to be tampered with. Those struck by calamity must have sinned and deserved the punishment. If some are masters and others slaves, so be it. Nothing could better justify the hypocrisy of American exceptionalism than Leibniz's naïve optimism that every event, however unpleasant, must have its rightful place in God's master scheme. In the modern Republican Party version, we don't have to worry about climate change because (in this best of all worlds) God will come to the rescue and we don't have to correct inequality or provide medical coverage because it's God's will.

Jonathan Swift wrote *Gulliver's Travels* two hundred years after More's *Utopia,* one hundred years after Shakespeare's *Tempest,* fifteen years after Leibniz's "Best of All Possible Worlds," seven years after Daniel Defoe's optimistic *Robinson Crusoe,* and fifty years before the Declaration of Independence. It is a wickedly funny, mordantly biting refutation of all positive appraisals of the human condition. Gulliver begins his travels heading for Bermuda, the focus of so many utopian dreams and dystopian nightmares. He lives up to his name by being gullible, affably naïve, and affectionately

interested in his fellow creatures. He ends his travels with such mis-
anthropic disgust toward men that he can no longer stand the sight,
sound, or smell of them. His journeys to strange places have shown
him the folly, pettiness, puffery, deceit, self-deception, selfishness,
indifference, and viciousness of man. The more optimistically we
start life, the more we will be disappointed by it. The more unre-
alistically optimistic our hopes, the deeper into pessimism will be
the plunge caused by bitter experience.[7]

Voltaire was next in the line of hysterically funny dystopians,
taking rare delight as he chopped Leibniz's utopianism into even
finer sarcastic mincemeat. In *Candide, or, Optimism,* naïve Profes-
sor Pangloss stubbornly persists in reassuring naïve student Can-
dide that "all is for the best in this the best of all possible worlds"
despite their repeated sufferings from one dreadful experience
after another—war, disease, famine, fire, flood, earthquake, hu-
man betrayal, deceit, and hypocrisy. A realist might conclude that
this is the worst of all possible worlds, but Pangloss's blindly opti-
mistic bias is undiminished. Voltaire instructs us that only willful
blindness to empirical experience can disguise that our world is
already very far from perfect and can always get much worse.[8]

The Actual American Experience

The essence of the American dream is aspiration, both on the
individual and the collective level. We are a country built by im-
migrants coming from a worn-out, overpopulated, strife-ridden
world that had spit them out, into a new and welcoming land
that at least gave lip service to the ideals of liberty, equal oppor-
tunity, and success through hard work. But aspiration has not
been the equivalent of actualization. Our Declaration of Inde-
pendence, stating that "all men are created equal," provided great
inspiration—but 240 years later has yet to become reality. "Amer-
ica" remains an ideal more than a reality—a noble enterprise, still

in progress and the source of justifiable pride, but also, alas, of much disillusionment.

More's imaginary American Utopia preceded by almost a century the settlement of the first permanent English colony on American soil—and played a part in influencing its direction. His reforming impulse (filtered through Calvin and the Quakers) inspired the British colonists as they set about creating what they hoped would be their new earthly paradise. The colonization of the New World by the Old is often romanticized (and Disneyfied) as an idealistic search for religious freedom and political perfection. From the very start, American utopianism and idealism had to compete with hardheaded American commercialism. The less exalted motives—land hunger, business opportunity, finding a place for second and third sons, running from the law—are glossed over because they don't offer a founding myth inspiring enough to support and justify American exceptionalism.

But we shouldn't swing too far in the opposite, revisionist direction and underestimate the sincerity of the utopian dream. Many who took the risky journey across the sea and onto barren beaches did so with the sincere hope of creating a much better new world once they were freed from the dispiriting institutions, relationships, and traditions that had anchored mankind in the depressing traditions of the old. The unpopulated, virgin land ("Indians" being conveniently repressed from the reforming colonial consciousness) offered a blank slate to correct the many evils mankind had accumulated in Europe. A fresh start for each individual man and woman; for mankind, a second chance at redemption.

The Mayflower Compact, signed just before the Plymouth landing in 1620, was an idealistic solution to nagging practical problems. The *Mayflower*'s occupants were divided about equally between religious dissenters (calling themselves the "Saints") and commercial opportunists (called by them the "Strangers"). Aware that internal dissent was an unaffordable luxury given the exigencies of their parlous external circumstances, Saints and Strangers composed a compact to "covenant and combine our selves together into a civil body politic . . . to enact, constitute, and frame such just and equal laws, ordinances, acts, constitutions and offices, from time to time, as shall be thought most meet and convenient for the general good of the Colony, unto which we promise all due submission and obedience."[9] This social contract resolved their different motivations through democratic self-governance by common consent for the common good.

Ten years later (also on a ship in the moments just before landing, this time in what would become known as Boston Harbor), John Winthrop preached a sermon, "A Model of Christian Charity," intended to set the tone for the new Massachusetts Bay Colony. He quoted Jesus's Sermon on the Mount, enjoining fellow Puritan colonists to create "the light of the world" and invoked St. Augustine's injunction to build "the shining city on a hill." Winthrop realistically warns his brethren that life in the New World will be hard and that its rewards will not be equally distributed. "God Almighty in his most holy and wise providence, hath so disposed of the condition of mankind, as in all times some must be rich, some poor, some high and eminent in power and dignity; others mean and in subjection." But despite their different backgrounds, skills, and wealth, people must live in interdependence as if different parts of one body, with the kind of love and loyalty that binds mother and child. Community needs must take precedence over individual need. All must work together to create a better world and to become a model for others to follow.[10]

It didn't quite work out that way. The colonists could not transcend the Old World as they created the New. Given an opportunity to forge a brave new world, they instead fell prey to the same psychological and social forces that follow mankind however far we roam. Even with the conscious intent, and the religious injunction, to behave justly, the colonists often did wrong. From its beginning, the Massachusetts colony was riven by strife and infamous for intolerance, prejudice, and superstition—hanging religious and political dissidents, grabbing land, killing Native

Americans, and conducting the Salem witch trials. It did not create a better world, nor was it a very good model for the existing one. The circle was vicious—utopian ideals degenerating into societal delusions, justifying horrible acts, betraying the utopian ideals.

Lest we fear that history must always repeat itself in the most negative way, a parallel experiment in government, based on different, more realistic principles, had a very much more favorable outcome. Seven years after the Massachusetts colony was established, its leaders banished Roger Williams as an unwelcome dissident to its intolerant practices and hypocritical idealism. He and some like-minded followers went into the wilderness—where they founded Providence Plantation, which in 1663 became part of the Colony of Rhode Island—on assumptions quite opposite to those he detested in Massachusetts. Williams had a common-sense appreciation of human psychology and a healthy skepticism concerning our ability to understand God's will. He feared the corruption that inevitably infects any group able to use religious authority to justify self-interest.

Realism turned out to be a much more humane and effective guide to successful governance than the exalted claims of idealism, exceptionalism, and religion. Williams established governance by and for the people, serving their secular aims rather than modeling a religious paradise. Leadership was by consent and based on an infrastructure of practical political institutions, not preordained by the self-serving and wish-fulfilling Calvinist myth that power came from providence. Williams introduced the then-novel idea of a "wall of separation" between church and state to guarantee absolute religious liberty. The colony quickly attracted an enterprising mix of Baptists, Quakers, Jews, and other religious minorities in what became the first American melting pot. Williams had the then equally strange notion that Native Americans were people—and were therefore automatically endowed with the human rights to hold land and live free. He purchased rather than expropriated their property and dealt with them by treaty, not fiat. Williams understood the importance of creating workable, everyday, political solutions and relationships—rather than trusting to the blindly optimistic and hypocritical utopian conception that we can create God's City upon the Hill.[11]

Forty years later, the new Quaker colony in Pennsylvania undertook a "holy experiment" to further consolidate a practical, rather than utopian, approach to governance. Its "Frame of Government" established the constitutional tradition that rule of law would protect individual rights—freedom of worship, free elections, fair jury trials, protection of the individual from arbitrary state power, impeachment of miscreant public officials, and restriction of capital punishment.[12] The other colonies were much quicker than Rhode Island and Pennsylvania to mouth idealistic religious aspirations, but much less likely to practice good works.

The first, and most inspiring, words of our founding document, the Declaration of Independence, read: "We hold these truths to be self-evident, that all men are created equal, that they are endowed by their creator with certain unalienable rights, that among these are life, liberty, and the pursuit of happiness." In writing the Declaration, Thomas Jefferson was strongly influenced by More's *Utopia,* and he was familiar with Leibniz's idealism. But Jefferson also owned slaves and must have realized that his reality did not come close to measuring up to his ideals. It was anything but "self-evident" in slaveholding America that all men are "created equal." There was also nothing in his personal experience, or in the experience of the new nation he was declaring, to suggest that all people are born with "unalienable rights"—his slaves' rights were badly alienated in Jefferson's own beloved Monticello. The United States was born with lofty utopian ideals that were daily contradicted by its grim daily realities, a split that runs through our history and remains apparent to anyone reading today's morning newspaper. The good news was that no nation had ever before promised its people (or, more precisely, those not enslaved) the individual right to "life, liberty, and the pursuit of happiness." The bad news was that the new nation often didn't keep its promises.[13]

The misinterpretation of the phrase "the pursuit of happiness" has also cheapened the currency of American exceptionalism. Jefferson was borrowing words from the philosopher John Locke, who, one hundred years before, had written that "no one ought to harm another in his life, health, liberty or possessions" and "the necessity of pursuing happiness [is] the foundation of liberty."[14] The term "happiness" had a special technical meaning for Locke and Jefferson that is quite different from modern hedonic connotations. For them, pursuing happiness meant becoming a better person and a more responsible citizen. They used "happiness" in the Greek philosophical tradition where it referred to civic virtues of courage, moderation, and justice, not to individual pleasure or enjoyment. In his *Nicomachean Ethics,* Aristotle wrote, "The happy man lives well and does well; for we have practically defined happiness as a sort of good life and good action."[15] Locke made this explicit: "We are, by the necessity of preferring and pursuing true happiness as our greatest good, obliged to suspend the satisfaction of our desires in particular." Illusory happiness is not "true and solid" happiness.

The pursuit of happiness was included in our Declaration of Independence and is a "foundation of liberty" precisely because it was meant to make us better citizens freed from enslavement by individual desires. As Jefferson put it, "Our greatest happiness does not depend on the condition of life in which chance has placed us, but is always the result of a good conscience, good health, occupation, and freedom in all just pursuits."[16] Americans have ever since vigorously pursued happiness—but too often we have chased after the easy vision of happiness peddled by Madison Avenue, not at all the civic virtue Aristotle, Locke, and Jefferson had in mind. The ever-practical Benjamin Franklin could see this coming: "The Constitution only gives the right to pursue happiness. You have to catch it yourself." Soon we will be discussing how we can best pursue real happiness in a sustainable world, not the fake and transient consumer pleasures that often govern ours.

Lincoln best embodied a more elevated, aspirational dimension of American exceptionalism (though he was fully aware that we were just a work in progress). It was not enough that we develop our own exemplary way of life. Providence demanded we light a beacon for a better world. His Gettysburg Address included the resolve that "these dead shall not have died in vain—that this nation, under God, shall have a new birth of freedom—and that government of the people, by the people, for the people, shall not perish from the earth." The irony was certainly not lost on Lincoln, as he delivered this speech in November 1863 on a blood-stained battlefield, that the United States was fighting the most brutal of civil wars and for all the wrong reasons—that we were, at that precise moment, perhaps the worst possible model for a better world. But Lincoln never gave up the hope that once the states were reunited, they would eventually heal wounds, regain the moral high ground, and lead others unto redemption. Lincoln was a secular preacher, always aware of the tragic flaws in mankind and in America, but always fishing for, and often finding, our better angels. If we are a chosen people, we must choose the right—to become "exceptional" in our goodness, not in our greed.

"The past is never dead. It's not even past."[17] The racism that condoned slavery never died, it just took on subtler hues. Reconstruction might have realized Lincoln's vision for a more just America had he survived to guide it. But the flawed men who followed him undermined emancipation—and the dreadful consequences are still apparent in today's pervasive racism. Although blacks were freed on paper 155 years ago, they were subjected first to the violence and imprisonment of strict Jim Crow apartheid and now must live within a system of shameful racial and economic injustice. America's greatest book (*The Adventures of Huckleberry Finn*) by America's greatest writer (Mark Twain) punctured white hypocrisy by showing that "black lives matter."[18] But America's first great movie (*The Birth of a Nation*) exalted the virtue of the

Ku Klux Klan and Trump won the election with its enthusiastic support. The hypocrisy of the Declaration of Independence ("all men are created equal," except slaves) has been replaced by the daily hypocrisy of black lives that are often separate, almost always unequal, and that don't matter nearly enough. The Civil War never ended—the Confederacy having won its latest battle by electing Trump.

Mark Twain also hated American imperialism, hiding under pious hypocrisies like "Manifest Destiny" and "civilizing mission." All men may be created equal, but American men are providentially endowed (or self-appointed) to conquer other men—to decimate Native Americans blocking their march west; to defeat Mexicans in a major land grab; and to win colonies in a manufactured war with Spain. From Jackson, to Polk, to Teddy Roosevelt, to Bush, we have had a succession of presidents willing to exert American power to push the envelope of our ambitions. Mark Twain saw Theodore Roosevelt "as the most formidable disaster that has befallen the country since the Civil War" and joked bitterly that "God created war so that Americans would learn geography." He could no more understand why the United States was killing people in the Philippines than I can understand why we killed people in Vietnam, or in the never-ending wars in Afghanistan and Iraq. American exceptionalism made all our wars seem perfectly just to us, even when they were patently unjust to the people we so blithely invaded. We often don't see things as they are; instead we see through the lens of our commercial interest and cloak our greed with a thin patina of idealism.[19]

Electing Entertaining Exceptionalists

In 1776, the combined population of the thirteen colonies, soon to become the United States, was only 2.5 million—among whom only a tiny fraction were adult, propertied white males eligible to vote and to hold political office. From this tiny pool emerged George Washington, Thomas Jefferson, Benjamin Franklin, Alexander Hamilton, James Madison, James Monroe, John Adams, John Jay, and dozens of other highly distinguished men. Our Founding Fathers were intellectuals of the highest order, well versed in enlightenment political theory, history, philosophy, science, economics, and rhetoric. The *Federalist Papers* are classics of political philosophy and our constitution has withstood the trying test of time. Fast-forward almost 250 years—to the reality that our population has increased 150-fold, but our political talent has decreased by what may be an equal proportion. Very few of our subsequent politicians have claimed anything close to the stature of our first-team Founding Fathers.

And of late, glitz seems to have gained a big edge on greatness. Every politician has always been something of an entertainer, but only recently have we thought it reasonable to turn to the entertainment industry to provide our presidents—Hollywood gave us Reagan and reality TV gave us Trump. Who better to weave our societal delusions than a well-practiced professional in the art of spinning fantasies and disguising truths. Hollywood thrives on drama, human interest, sentiment, and the eternal conflict between good guys and bad guys—with good guys always winning and bad guys always punished. Simple scenarios that avoid the wrenching complexity and necessary compromises of real life. Most commercially successful movies, and all successful reality TV, achieve popularity precisely because they don't require too much thinking. Ronald Reagan was the most gifted politician of his time, in large part because he was so good at acting the role of president, unencumbered by thoughts about the many harmful unintended consequences of his mistakes. Donald Trump became president in large part because reality TV had made trash-talking a political asset. We have fallen far from the proud but humble patriotism of Washington and Jefferson to the "America Is Great" bombast of Reagan and Trump. Righteous David has grown up to become a swaggering Goliath.

Everybody's All-American

Reagan was an almost pure expression of all that is wonderful and terrible in American exceptionalism. Born in modest circumstances and into a troubled family, he overcame all obstacles to rise to the top of America's two most glamorous professions—Hollywood movie stardom and Washington political power. Admired by some as a great president and derided by others as one of our worst, he was in fact both. Reagan's infectious optimism and "don't worry, get happy" message worked a miracle of national cheerleading and morale building. He took over a country stuck in the malaise caused by Jimmy Carter's truth-telling pessimism and immediately cheered us up. Reagan sold a rose-colored version of America that was every day fulfilling John Winthrop's hopes—we were the precious reincarnation of the "shining city on a hill," the envy and the admiration of the entire world.

Reagan was a master at creating illusions and even better at selling them—perhaps because he was such a convinced believer in his own myths. When his movie career fizzled, he got a job hosting a western TV series where he perfected pitchman skills selling soap products. He also hosted a long-running television drama series for General Electric, successfully convincing viewers that a happy life depends on owning every conceivable electrical appliance and gadget. Reagan became the symbol of modern American consumerism and the most persuasive promoter of our wasteful energy usage.

The other half of Reagan's GE gig was seemingly prosaic, totally unglamorous, but turned out to be far more important—dramatically changing his own personal politics and, soon after, the politics of the entire nation. During the period between 1954 and 1962, Reagan spent about one-fourth of his life on the road giving inspiring speeches to hundreds of thousands of workers at 139 GE facilities and tirelessly spreading the GE gospel at countless Chamber of Commerce and Rotary Club dinners, in small towns all across the United States. Here Reagan was pitching the model of the world dreamed up by the GE public relations department. And when you pitch a new view of life that long and that hard, you wind up selling yourself on it—especially if you are a great salesman like Reagan. He had started his long GE travels a fairly liberal Democrat—he ended swinging so far to the political right of the GE corporate dogma that the company could no longer use him. He may have started his speechifying as a decidedly mediocre movie and TV performer, but he ended it the most accomplished political speaker since Franklin Delano Roosevelt. Reagan was transformed into the "Great Communicator" through years of on-the-job GE training, sometimes giving fourteen speeches a day, and well into the night. He knew what the American people wanted to hear and developed a folksy, easy way of saying his lines and selling his political goods. Reagan rapidly transformed himself from GE shill to rising Republican star— giving a celebrated speech at the 1964 Republican National Convention, becoming governor of California from 1967 to 1975, and president of the United States from 1981 to 1989.

We are still paying the price for Reagan's feckless presidency and promotion of societal delusions. His supply-side "voodoo" economics tripled our national debt. He assured Americans that we could afford to live large—we should love our big energy-guzzling cars and houses, not worry about cost, waste, or environmental impact. His beaming smile and cheery countenance disguised the fact that he was massively redistributing wealth in just the wrong direction—to the rich and superrich much must be given, because perhaps a small bit of it may someday trickle down to the rest of us. Deregulating the banks helped promote fiscal shenanigans that led to financial meltdowns. Deregulating industry led to pollution, monopoly pricing, and industrial accidents. Reagan's naively optimistic foreign ventures had even more costly consequences. Supporting Islamic "freedom fighters" against Russia in Afghanistan backfired when they became Islamic "terrorists" and began turning our own weapons against us. Collaborating with Pakistan helped further our "great game" in Afghanistan, but facilitated a Pakistani nuclear weapons program that is now dangerously close to falling into the hands of terrorists. Reagan financed (often illegally) right-wing insurrections in Latin America that enhanced the continent's enduring anti-Americanism. And he did secret, dirty deals with Iran that called our national integrity into serious question.

The country fell in love with Reagan in a big way. The unrealistic optimism of American exceptionalism feels great while it lasts, but it can never last very long and left us with a big hangover. Sooner rather than later, reality bites and all the accumulated debts come due. Thirty years later, we are still stuck with the bad hand Reagan dealt us.

Trump Exceptionalism

A portion of our country also managed to fall deeply in love with Donald Trump—a "reality" TV star who maintains an even more tenuous relationship with reality than did his hero, the much more lovable Ronald Reagan. Trump and Reagan have opposite temperaments—Reagan eternally sunny, Trump ever dark, but both sell similar societal delusions. Trump manages to be truculently wrong on every existential question facing humanity—denying global warming, encouraging pollution, promoting resource depletion, enjoying saber rattling, opposing population and gun control, escalating obscene inequality, and trampling on civil rights protections. He has the demeanor of a circus barker, the integrity of a con man, the temperament of a neighborhood bully, the breathtaking ignorance of an arrogant know-nothing, the political instincts of a führer, and the policies of a tribal nationalist.

All the problems that are inherent in American exceptionalism are now very badly compounded by Trump exceptionalism. Trump doesn't qualify for a mental disorder, but he does present with one of the world's best-documented cases of lifelong failure to mature. He is a boy/man who expects everything to go his way and experiences the world as an extension of himself. Other people exist only to do his imperious bidding, admire his great deeds, and gratify his enormous wants. This is perfectly age-appropriate behavior in a young child, but is perfectly inappropriate in a president. Trump fancies himself the nation's Big Brother, but he is really our neediest Big Baby. Trump is bad, not mad, but we the people are mad for having elected such a terribly flawed person to the most powerful position in the world. His heady rise was surprisingly undeterred by his consistent pattern of bold-faced lying; constant flip-flopping; irresponsible incitement to violence; and self-congratulatory bigotry, racism, and sexism. Blustering, bullying, and bravado play well on reality TV but can be disastrous in real life when you are running a country. No one less qualified to be president has ever won the office. No one so dangerous to our democracy has ever been given its most powerful position.

Trump is an unlikely messenger bearing an unwelcome message about the sanity of our body politic. He has revealed and unleashed a deeper streak of delusional denial in a larger segment of U.S. society than even I would have thought possible. Thomas More's hope for a utopian America has decayed into a Trumpian dystopia. Trump's slogan "Make America Great Again" (stolen, appropriately enough, from Ronald Reagan) disguises policies that in fact make America small, fearful, angry, petty, and vicious. Karl Marx once quipped, "History repeats itself, first as tragedy, second as farce." The Trump phenomenon marks what I dearly hope will be a bottom in both American tragedy and farce.

Democracy is a precious, but historically infrequent and perilously fragile method of governance. Athens initiated democratic government, but its brief experiment ended in failure when the people were seduced by demagogues into disastrous decisions. Plato believed democracy to be so unworkable an institution that he banned it from his ideal Republic. When forerunners of Western democracy began emerging four hundred years ago, the philosophers Hobbes and Vico predicted they would inevitably lead to chaos and a return to all-powerful central control. The last three hundred years have proved democracy to be the best of governments when it works well, the worst when bedeviled by divisiveness, disorganization, chaos, and corruption. The world now contains dozens of failed "democracies" in the midst of, or verging toward, civil war, anarchy, and/or totalitarian takeover. There is an old Islamic quotation: "Better one hundred years of the Sultan's tyranny than one year of people's tyranny over each other."

Apart from our own brutal Civil War, U.S. democracy has seemed a charmed bedrock of stability—surviving mass migrations, periodic depressions, and stark economic inequality. But the future holds no guarantees. When, forty years ago, Henry Kissinger began making small talk on first meeting Zhou Enlai, he asked his opinion on the French Revolution. Zhou replied: "Too soon to tell." Likewise, it is too soon to tell whether American democracy can survive Trump's attack on it. He may be no more than a blowhard and buffoon, but Trump has proven to be no joke. He has embodied and unleashed forces that seriously undermine our democratic principles and institutions. Because of Trump, the Freedom House 2017 ranking of democracy has dropped the United States to the thirtieth in the world and has us trending downward.

Trump markets himself as the law-and-order candidate but has displayed a regal contempt for law whenever it contradicts his interests, impulses, whims, and grudges. He feels entitled to squash critical press, throttle judges who don't see things his way, force the military to torture in violation of international law, and break treaty obligations. Trump doesn't understand or respect the delicate balances built into our constitution and feels no compunction whatsoever in distorting them out of all recognition. Trump says or does at least one obnoxiously autocratic thing almost every day, but never has to pay the expected political or personal price. He once bragged, probably accurately, that he could commit murder in broad daylight and not lose support. It is but few steps from this attitude to an attempt at dictatorship.

To believe that our democracy is secure forever requires blindness to the current resurgence of antidemocratic trends in most of the world's democracies, including our own. Fueled by fear, uncertainty, nationalism, economic distress, xenophobia, and racism, right-wing radical parties and policies are quickly gaining votes, traction, and respectability. The reflexive overresponse to terrorism has been to diminish civil rights and ramp up surveillance. Democracies historically fail when they make incompetent decisions, or suffer from a paralysis of indecision, leading to chaos and hostile takeover by a strongman. Trump's governance has put on full display not only his own breathtaking bias and ignorance, but also the blatant unpreparedness and incompetence of his sycophantic cabinet cronies, who are all too eager to bow to his worst whims and support his most unsupportable prejudices. Our country already suffers from a widespread distrust in government. The burlesque disorganization of the Trump administration may realize the worst fears of many that democratic government no longer works, creating a vacuum to be filled by someone who can make the trains run on time.

There are precious few institutional obstacles to a complete Trump takeover. The Republican Congress daily abdicates what should be its primary patriotic duty (providing a check on Trump's autocratic ambitions) and instead cynically chooses to use him as facilitator of its right-wing agenda. Not surprisingly, Trump's most furious attacks are aimed at the two remaining bastions of American democracy—its free press and its justice-protecting courts. The most frightening of all Trump's frightening tweets: "The FAKE NEWS media (failing @nytimes, @NBCNews, @ABC, @CBS, @CNN) is not my enemy, it is the enemy of the American People!"[20] Trump sets himself up as defender of the American people as he mounts the charge against their most fundamental freedoms of speech and thought.

Trump's second-scariest tweet was "Just cannot believe a judge would put our country in such peril. If something happens blame him and court system. People pouring in. Bad!"[21] He was reacting to court decisions questioning the constitutionality of his banning travelers from Muslim countries and, even more important, his claim that the courts have no jurisdiction on his acts because he is protecting the higher value of national security. Trump has handed himself a win-win situation. If the courts decide in his favor, he can use the national security excuse to assume dictatorial powers. If the courts provide a check on his autocratic disrespect for the Constitution, they (not he) are to be blamed for any terrorist act—and he can respond by claiming emergency powers. Let's hope we never see how the military would respond to such a Big Brother Trump order. It certainly can happen here—we are now vulnerable to the imperial ambitions of an impulsive, rabble-rousing con man who does not feel bound by the checks and balances built into our constitutional limitations.

Trump is our first Demagogue-in-Chief, a man with no respect for American democratic institutions. His administration is considering new legislation that would liberalize libel laws to muzzle the press and change congressional procedures to reduce the checks and balances that protect democracy from becoming tyranny. It is especially scary that Trump never met an autocrat he didn't like and want to emulate—Putin of Russia, Erdogan of Turkey, el-Sisi of Egypt, Modi of India, Duterte of the Philippines. He supported the anti-Semitic and neo-fascist Marine Le Pen of France and even has kind words to say about Kim Jong Un of North Korea.

Many people fret about Trump's motivations, psychology, and possible psychiatric diagnosis. I think these are irrelevant. It doesn't matter that much whether Trump is crazy, crazy like a fox, or just an incompetent boob who keeps getting lucky—or some combination of all three. What does matter greatly is stopping him—and stopping him now before it is too late. This is a tipping point, a time of trial for the soul of America—we will protect our democracy from Trump's frontal assault or we will lose it. My friends teased me that I was crying wolf about the risks of an American brand of fascism when the dark Bush/Cheney team exerted executive power in ominous ways in 2000. Now those friends are as scared as I am. More tellingly, Norman Ornstein, the wisest and most objective political observer in Washington, is also scared. "We don't have a conventional president. We're seeing behavior that could lead us right down the path to martial law or authoritarian rule." We must stand up to Trump or he will trump our democracy.

Being Exceptional Hasn't Always Meant Being Good

Among many Americans in the dark age of Trump, "American exceptionalism" has increasingly degenerated into self-congratulatory sneering that we are better, more righteous, wiser, more idealistic, stronger, smarter, kinder, fairer, less selfish, harder working, and more generous than other nations. The United States is portrayed as the greatest—the land of opportunity with the best of all possible economic systems, the best political institutions, the best people, the best intentions. There are also religious overtones—we are God's country, providence is on our side, and we are leading the way to a Christian millennium. Some of the brags have been true, but only some of the time. By luck, pluck, ambition, ruthlessness, resources, and historical accident, American exceptionalism became a self-fulfilling prophecy. The United States could achieve its self-declared "manifest destiny" by buying and conquering almost an entire continent and soon became the richest, most technologically advanced, and most powerful country in the world. But there were always troubling negatives embedded within the positives. To achieve economic, military, and political greatness, the United States often found itself consciously or unconsciously practicing exactly the opposite of what it was preaching.

American exceptionalism also enables societal delusions that blind us to reality and distort our responses to existential crises.

Assuming we are wiser than other countries prevents us from learning from them. Assuming we are mightier than we are encourages us in the futile attempt to police the world. Assuming we are morally superior leads to puzzlement when others so often see us so negatively. Assuming we are self-sufficient prevents us from cooperating with other countries on worldwide problems, whose solutions can only come through collective action. Triumphalist, tribalistic, neoconservative nationalism has created the most nakedly self-destructive expressions of modern American exceptionalism. We fight only just wars, only for unselfish reasons, and always fight them fairly—ergo, any war we happen to be fighting must necessarily be just, unselfish, and fair. We have a civilizing mission to be the world's teacher, preacher, and policeman. People in other countries appreciate our unselfish altruism and welcome our interventions. The (temporary) end of the Cold War was proof of our righteousness and of our right to tell others how to live. We must create an "American Empire," based on spreading democracy and globalized free trade to the farthest reaches of the earth.

This "exceptionalist" delusion has gotten us into lots of trouble. Viewing the world in us-versus-them, good-versus-evil, Manichean terms has led into a repetitive series of depressing foreign policy disasters across times, administrations, and parties. Our popularity is in steady decline. After World War II, surveys showed we were at the very top of the world's list of admired countries, and Germany was at the bottom—now our rankings are reversed. A recent BBC poll of 24,000 people in 33 countries found us to be the second most unpopular country in the world—ahead of Iran, but less popular even than some pretty unpopular places, like Russia, Saudi Arabia, Zimbabwe, and China.[22] A Gallup poll of more than 66,000 people in 65 countries asked which country "is the greatest threat to peace in the world today." The United States topped the list at 24 percent, followed by Pakistan (8 percent); China (6 percent); and then a tied score for fourth place featuring Afghanistan, Iran, Israel, and North Korea, all with 4 percent. Some of our bad repute is the inevitable result of being the world's only superpower—big guys naturally have enemies and stimulate envy even among friends. But a lot of the anger toward us comes from the fact that we haven't been a very smart superpower. Sixty years of interfering in the internal affairs of other countries, toppling democratically elected leaders, encouraging right-wing rebellions, starting dumb and destructive wars, spreading disinformation, resorting to torture, and exploiting countries' other resources have taken an understandable, but regrettable, toll.[23]

Narcissism, in a person or a country, is adaptive in small doses but disastrous in large. In moderation, national narcissism allows us to feel confidence in ourselves, inspires others to be confident in us, and promotes decisive decision making and vigorous action. But overweening pride usually goeth before a great fall—both in an individual and in a country. American exceptionalism, once a great national asset, has now become an equally great national liability that renders us particularly susceptible to forming comfortable, but dangerous, societal delusions. All the causes of bad decisions discussed in the previous chapter are amplified whenever our narcissism obscures our reality, clouds our judgment, and lends a false aura of righteousness and wisdom to what are sometimes very dumb actions and inactions. Teenagers usually outgrow their age-appropriate narcissism and develop into adults. We are still a teenage country, with relatively little real-life experience and a lot of growing up to do. More adult countries, having existed through many cycles of good times and bad, have always regarded our exuberance with indulgent apprehension—now converted to sickened amazement at the spectacle of Trump. We must mature as a nation if we are to shed societal delusions, provide leadership to the world, and grow into a comfortable middle and old age. This certainly won't be easy (and recent Trumpian trends are an absolutely disastrous regression), but necessity is the mother of invention and the United States has always been pragmatically clever at reinventing itself. Perhaps we will not only endure Trumpism, but prevail over it.

If we are to work and play well with others, we have first to be clear-eyed about how the outside world sees us and why. Having the utopian delusion that we are a shining example with unique wisdom prevents us from promoting policies that would earn respect, rather than fear and anger. The particular circumstances of our history created institutions that have worked more or less well for us, but do not translate seamlessly to peoples with other histories, other cultures, and other aspirations. We would have a more positive impact on the world, and gain better standing and reputation, if we were humbler in our goals, methods, and claims. There is no serious world problem that can't be made much worse if we blunder naively into thinking we can easily and uniquely make it much better.

Dystopias Become Bestsellers

Dystopias have become a dime a dozen, almost a cliché. The genre began to expand exponentially with the beginning of the Industrial Revolution, exploded further when robots entered the picture, and took over the popular imagination when the problems of the modern world began overwhelming the solutions and the future started looking bleaker than the past. Dystopias now cover different age groups (kiddie, teenage, young adult, adult); have different breeds of heroes and villains (human, animal, robot, computer, alien, hybrid); occur in different eras (past, present, and especially the future); come in different genres (romance, fantasy, satire, melodrama, comedy, tragedy, horror, and especially science fiction); focus on different themes (war, totalitarian government, revolution, anarchy, imprisonment, surveillance, psychological torture, environmental catastrophe, class struggle, caste systems, tribalism, overpopulation, resource depletion, pollution, crime, cannibalism, cloning, economic disaster, political intrigue, societal breakdown, cultural decline, the collapse of the family, dehumanization, loss of identity, industrialization, fanaticism, religious cults, the recursive cycles of history, and lately especially the impact of new technologies); and achieve different levels of excellence (trash, lowbrow kitsch, middlebrow kitsch, great literature).

The three most directly relevant modern dystopias are Aldous Huxley's *Brave New World* (a scary extrapolation of capitalist America),[24] George Orwell's *1984* (a scary extrapolation of communist Russia),[25] and Sinclair Lewis's *It Can't Happen Here*.[26]

Huxley and Orwell were both remarkably accurate in their predictions of our modern world. Orwell's book is far greater as literature; Huxley's a more accurate rendering of the United States (at least as it is today—who can predict the future given Trump's Orwellianism). The only things Huxley got wrong are the location of power and the motivating force that has driven the creation of "Our Brave New World." He predicted that government would use hedonic manipulation to exert political control. So far, at least, it has instead been corporations manipulating hedonic pleasure to extract profit. But these are still early days—too early to rule out government as a future threat, especially with Big Brother Trump breathing down our necks.

Enter Huxley's *Brave New World*

In some ways, it will seem quite familiar since it was an American dream/nightmare world modeled on Huxley's adopted hometown, Hollywood. Imagine a society in which the Golden Rule is "Never put off till tomorrow fun you can have today." You are encouraged to purchase whatever you like the moment you feel the impulse. Promiscuity is moral and monogamy is pornographic. Guilt-free sex is available with whoever catches your fancy, but falling in love is regarded with disapproval and disgust. Your job is perfectly tailored to your abilities by genetics, cloning, and behavioral conditioning. Everyone around you is friendly, cowlike, compatible, and shares your banal tastes and interests. You have no bad memories from the past; no problems in the present; no worries about the future. "People are happy; they get what they want and they never want what they can't get. They're well off, they're safe; they're never ill; they're not afraid of death; they're blissfully ignorant of passion and old age."

If anyone in *Brave New World* ever feels even the slightest twitch of sadness or anxiety, there is Soma, the magical happy pill. "There's always Soma to give you a holiday from the facts . . . to calm your anger, to reconcile you to your enemies, to make you patient and long-suffering. You swallow two or three half-gram tablets and there you are. Anybody can be virtuous now." You live in a stable, well-ordered, strifeless state that has perfected the governing principle that the people must not "lose their faith in happiness as the Sovereign Good and take to believing instead . . . that the purpose of life was not the maintenance of well-being but some refining of consciousness, some enlargement of knowledge." Reproduction is no longer connected with love or sexual attraction and is no longer subject to the uncertainties of fickle fate. Science has bypassed intercourse and pregnancy as intermediary steps and instead produces preplanned test-tube babies hatched in incubators. Cloning is so efficient that one egg fertilized in vitro can produce ninety-six identical twins. A well-run assembly line can turn the eggs of one woman into "an average of nearly eleven thousand brothers and sisters in a hundred and fifty batches of identical twins, all within two years of the same age"—not so very far from the studies in genetic engineering that are the hottest thing in science today.

There is a completely rigid, but comfortable, class system— genius alphas at the top, semi-moron epsilons at the bottom, and well-differentiated steps in between. People are all perfectly adapted to, and happy with, the varying expectations of each of their different lines and levels of work. Characteristics of each class are manipulated by genetic selection; exposure of the incubating fetuses to different chemicals; and constant, night-and-day behavioral conditioning specifically geared to prepare them for their predestined role. Children are raised in a "hatchery" that precisely controls and monitors every stage of their development.

The Malthusian population problem has finally been solved by uncoupling reproduction and sexuality and by removing all personal choice. Everyone loves having loveless sex and feels only disgust at any mention of old-fashioned habits like children being conceived via intercourse, or being born from a mother, or being raised by one's own parents. The few women allowed to reproduce have a "Malthusian belt," holding "the regulation supply of contraceptives" worn as a fashion accessory. World population is very strictly controlled at a constant and sustainable two billion.

Having promiscuous sex, great drugs, and a credit card with no limit would be for many the fulfillment of the American dream, but it was Huxley's idea of hell. His rebel-with-a-cause hero is a noble savage who protects his personal identity against state control and prefers Shakespeare to Soma. "I'm claiming the right to be unhappy. . . . I want poetry, I want real danger, I want freedom, I want goodness, I want sin. . . . I'd rather be myself and nasty, not someone else, however jolly. . . . I'd rather be unhappy than having the sort of false, lying happiness you were having here." For Huxley, unrestricted and mindless hedonic pleasure is incompatible with the creation of art, the progress of science, or the maintenance of human dignity. For humans to be completely happy in a hedonic sense, we would have to accept the bargain of becoming less than fully human.

Enter Orwell's World of *1984*

George Orwell's dystopian vision in *1984* presents an equally terrifying place, but in a very different way. Big Brother and the Thought Police watch your every move through Telescreens, hear your every word with hidden microphones, and have ubiquitous informers (including your children) to rat out your every thought, feeling, and relationship. The language is "Newspeak." In this looking-glass world, everything is the opposite of what it seems: The Ministry of Peace conducts perpetual war; the Ministry of Truth rewrites the past to conform with lying Party propaganda; the Ministry of Love does torture. The slogans of the day are "War Is Peace," "Freedom Is Slavery," and "Ignorance Is Strength." Failing to follow the party line is "thoughtcrime" and the good citizen has a deep "memory hole" to store dangerous and inconvenient truths. Those who have defied Party orthodoxy become "un-persons" to be written out of history. True is false and false is true. So very Trumpian.

All love must be reserved for Big Brother—individual unions are a crime against the state punishable with a very specific torture designed to press each person's most sensitive button. Knowing that Winston, the hero of the book, intensely fears and hates rodents, the Thought Police attach to his face a crowded cage of the largest, most ferocious and ravenous rats. The police won't tell him what he must say and feel to save himself, but just as the cage door is about to be opened, the right formula comes to him in a flash: "Do it to Julia"—the love of his life. His "spontaneous" betrayal of his loved one proves he has been cured of the insanity of individuality and can be welcomed back to society as a good and reliable citizen. Naturally, Julia was herself brought back to sanity by similarly betraying Winston. The Party wants not only their obedience, but also their love. The book ends with Winston crying into his gin, overcome with fondness as he sees Big Brother on the telly.

Until recently, Western readers of *1984* could feel a certain complacent superiority, convinced that its drab deceit, constant surveillance, and well-meaning cruelty were the special province of our enemies, especially Russia. We in the civilized world were pure and safe from totalitarianism. All this changed when Putin successfully used his KGB tricks to get Trump elected and Trump began copying the methods of his autocratic Russian tutor. Trump's every tweet and press conference is an exercise in Newspeak—shameless lies disguised as "alternate facts." Government websites that presented inconvenient truths have been wiped clean. Trump's biggest and most crucial battle is with the media—the most "dishonest" and "disgusting" people he has ever met. His fear, and consequent rage, are stoked by the free press's deeply held reverence for fact checking. Nothing is more dangerous to an autocrat than the simple truth. Nothing is more important to an autocratic government than to delegitimize the truth and the people courageous enough to tell it.

And even before Trump, Snowden's release of documents proved that the U.S. government had become a surveillance superstar snoop, that it has consistently lied to us, and that the CIA conducts psychological and physical torture not so different in technique and spirit from that performed by the Thought Police. Big Brother's tools for mind reading and thought reform were pathetically undeveloped compared to the technological snooping prowess now available to any would-be dictator. Privacy, free thought, and democracy have never been so vulnerable to autocratic manipulation.

Enter Sinclair Lewis's American Dictatorship

In Sinclair Lewis's still-chilling 1936 novel, *It Can't Happen Here,* Buzz Windrip, a skillful and charismatic demagogue, gains election to the U.S. presidency by making extravagant promises of miraculous economic gains, by stoking up the voters' anger and fears (not hard to do in the desperate fertile ground of a dreadful depression), and by appealing to patriotism, traditional American values, and hatred of Jews and foreigners. After election, Windrip takes on dictatorial powers, backed by a private militia that resembles Hitler's SS.

Lewis based Windrip on the character and aspirations of Huey Long, the homespun Louisiana demagogue of the Great Depression and the man in America's history most prefiguring Trump. Self-nicknamed "the Kingfish," Long ran on the slogan "Every Man a King." He had already asserted almost dictatorial power as governor of Louisiana and maintained it from afar, even after being elected to the U.S. Senate. Before his assassination in 1935, Long was Roosevelt's most hated and feared presidential rival (Roosevelt privately likened him to Hitler). Long's base was much better organized and proportionately much larger than

Trump's, with 7.5 million "Share Our Wealth" club members, a radio audience of 25 million, and 60,000 fan letters a week. Like Trump, he feasted on the adulation of crowds and the excitement of campaign rallies. Like Trump, he masqueraded as a man of the people in the service of personal ambition. Lewis played the counterfactual game of imagining what would happen here had Long not been assassinated and instead won the 1936 election against FDR. (Philip Roth mined the same territory in his *Plot Against America,* substituting a Lindbergh victory over FDR in 1940.) Lewis imagined his fictional fascist takeover of America by fusing Long's demagogic flirtations with the then-recent rise to power of fascist governments in Germany, Italy, and Spain.

It is no surprise that eighty years later *It Can't Happen Here* is back on the bestseller lists—life does sometimes follow art, just as art sometimes follows life. Trump is an almost exact, if more grotesque, replica of Buzz Windrip—and is the second coming of Huey Long.

Making America Great Again
Means Making It Good Again

I haven't given up on America. I love my country deeply, but often lately can't stomach what it says and deplore what it does. I am loyal to my country's noble values, to its breathtaking beauty, to its welcoming people, to its political institutions, to its history. I am forever grateful to America for saving my family of wandering Jews, who had lost almost everything and had no place else to go. America gave us a new lease on life, house and hearth, education, employment, a new culture and language, safety and security. Most important, America gave us optimism, something to believe in—an ideal of freedom in a world of doomed hopes. American exceptionalism was always somewhat accurate and often served our country well. We are unique in our combination of size, resources, wealth, commercial enterprise, productive capacity, and ability to attract and meld diverse waves of migration. And American exceptionalism has provided the ambition and optimism that made our country great and kept it unified (except for that one notable nineteenth-century exception). The New World rescued the Old in two world wars and provided it with a model of individual freedom and economic prosperity.

But it is past time to accept that American exceptionalism freights us with heavy baggage and is no longer an accurate reflection of our place in the world. Once responsible for half the world's productivity, we are now down to just 20 percent, still an enormous share, but certainly not dominant and smaller than both the European Union and China. And we can no longer afford the utopian delusions that allow us to avoid facing the realistic risks of the future. For the world to function well in meeting its challenges, the United States must accept that unilateral exceptionalism is self-defeating—bad for us, bad for the world.

We can learn how best to act cooperatively by following the example of our own national history. The post-Revolutionary "United States" was a disorganized mess until May 29, 1790, when the ratification of our constitution united us in fact, not just in name. We are now at a similar choice point—the world can't possibly respond to its existential challenges unless its individual countries create a really United Nations and it can't possibly solve its difficulties unless we solve ours. We are too big a part of the world's problems and too important a driver of solutions.

We are a country born in noble aspiration that collides with basic, less noble aspects of human nature. The gold of Lincoln, cheapened by the brass of Trump. It is great that we try; disappointing when so often we don't succeed. The American dream that started with such high hopes is now in its deepest hole. To me, real patriotism can never be "my country right or wrong." Slavish loyalty is much less loyal than clear-eyed and constructive criticism—failing to identify and correct our country's wrongs is to let them persist and fester.

Chapter 4

How Could a Trump Triumph?

*For every complex problem, there is an answer
that is clear, simple, and wrong.*
—H. L. MENCKEN

I used to be a compulsive news surfer and magazine junkie. That stopped sometime in early 2016 when Trump's ubiquitous media presence made me feel like screaming or retching. My anger and disgust went beyond his crude manners and egotistical self-promotion—although admittedly both were hard to stomach. Much worse was his cynical exploitation and promotion of all the prejudices and self-deceptions this book, which I was then writing, was meant to challenge. I had been hoping, in my very small and quiet way, to help make the world a bit more rational and cooperative. Trump had succeeded, in his very big and raucous way, in making it much crazier and more dangerous. I experienced Trump as a kind of secular antichrist, leading his supporters into an apocalypse of societal delusion.

Like many people, I had trouble understanding how so many Americans could possibly be so gullible. Wouldn't anyone in his right mind see through Trump's transparent lies and reckless flim-flammery? In our previous more reasonable world, Trump's many flirtations with running for the presidency had been discounted as a laughable circus sideshow. He was then seen for what he is: a would-be emperor with no clothes. But this go-round, Trump succeeded in mobilizing and exacerbating the denial, fears, and resentments (and sometimes also the hate, paranoia, racism, and misogyny) of people whose needs had been ignored by other politicians. People left out of the American dream were willing to accept Trump's simple and wrong solutions to complex problems. It was doubly depressing to watch the European Union do a similar slow-motion disintegration under pressure from so-called populist demagogues, exploiting the same insecurities and grievances. Britain was the first to fall with Brexit, and serious challenges to democracy and unity were mounted in the Netherlands, Austria, and France.

Then one evening, in the midst of what had become our customary dinnertime Trump lamentations, my wife called me out as a hypocrite. She said it was easy enough for us to feel superior to the Trump crowd, because we were so insulated from the problems and dangers they were facing. Trump might be the worst of all possible messengers, but that shouldn't allow us to discredit his message or disrespect those who were sending it. If Trump's supporters weren't suffering, they wouldn't be so susceptible to the blandishments of such an obviously false prophet. She reminded me of the great Eddie Murphy/Dan Aykroyd movie *Trading Places* (1983). Rich guy and poor guy are unexpectedly thrust into each other's role—the poor guy soon starts acting with the pretentions of the rich guy, while the rich guy goes into a poor guy's angry survival mode. Moral: you can't understand what the other guy is thinking and feeling until you walk many miles in his shoes.

Donna also pointed out what should have been perfectly obvious to me. I was writing a book preaching only to the choir, ignoring the beliefs and feelings of an important potential audience—the people who disagreed with us and supported, or even adored, Trump. She reminded me that you can only persuade people you like, understand, and respect. It was arrogant to feel superior to someone snookered by Trump—but for the grace of my unearned easy life, I too might well have become a Trump supporter. And it was misguided to think I could help people see through the societal delusions he was selling if I didn't understand why they needed to hold them.

As a psychotherapist, I should have known better. The best predictors of a successful treatment are therapist and patient liking each other and working well together. You don't have to like your patient from minute one. And you don't have to like everything about your patient. But to develop a healing alliance, you must eventually find enough redeeming virtues to balance any initial aversions. And it helps a lot to figure out precisely what you don't like about him and what in you makes him seem so bothersome. Most people become much more likable the better you get to know them—the more you understand their plight and your personal reactions to it. And the more people you get to know, the better and faster you are at picking up and taming your own prejudices. Eventually, most people turn out to be likable enough to work with. I can count on the fingers of one hand the number of people I couldn't treat because I could never find something valuable in them. But there were many hundreds I started off not particularly liking, but who I wound up liking a lot.

I realized immediately that I could never bring myself to like or respect Trump, even a small smidgen. I find him too deceitful and too self-serving to qualify for any benefit of the doubt. But my irrevocable loathing of Trump should not have spilled over to his supporters. Many millions of Americans are greatly burdened by our shameful inequality, stagnant standard of living, loss of good jobs, precarious medical coverage, porous safety net, and rapidly shifting cultural values. To write a meaningful book, I had to lose my biased snap judgment that they were all dumb or selfish or bigoted. Instead, I had to understand the problems, motivations, attitudes, and aspirations that attracted people to Trump. To write an effective book dispelling societal delusions, I had to overcome my disdain for the people who held them. To do this, I had to better understand them.

My new assignment was to flood myself with Trump and Trump supporters, not avoid them. I made a point of discussing Trump and his positions with as many people as I could find, starting with friends, family, and acquaintances and ending with anyone who was willing, which turned out to be most people. I also made myself spend endless hours watching Fox News, listening to right-wing talk radio, and following the Trump social networking machine. Some of it was just plain nuts—conspiracy theories driving a misogynist, racist, gun-toting mythology. The Internet, so promising a tool for creating an educated citizenry, also facilitates the spreading of the most blatant bullcrap. The low point in my fact finding and opinion canvassing was the declaration of a good friend—made with reddened and furious face, his jaw tightly clenched, an index finger pointed dramatically in the air—that he would vote for Hitler before allowing "that bitch" to be president. He lives in an alternate media universe from my customary "mainstream" diet, one in which Obama hates America and Trump will save it with his talent for "creative destruction."

Trump Exploits American Distress

Freud said that delusions don't occur in the sky—like dreams, they are a distorted expression of an underlying reality. You can't begin to treat a patient's delusion until you understand why he must believe it so strongly—the realities expressed in the delusion and his psychological reactions to them. Similarly, we will never cure societal delusions unless we understand the underlying problems promoting them and provide realistic solutions to replace wishful thinking. Trump is no more than a skilled snake-oil salesman selling quack medicine—but the societal sickness he is exploiting is all too real. He won power because he promised quick, phony cures for the following real problems burdening a significant segment of our population left out of the American dream.

Jobs

Between 1870 and 1970, the United States enjoyed the world's highest rate of wage and job growth. Early migrants to America had come in search of land; later migrants came in search of a good-paying job. But since 1970, the real wages in the United States have declined. An average worker used to be able to support his or her family on just one salary. Now both spouses must work at least one job—and it is still a scramble to make ends meet. It doesn't seem fair to hardworking, middle-aged whites without a college degree (Trump's most solid base) that they are worse off than their parents, especially since African-Americans and Latinos are now somewhat better off than theirs.

Obama inherited from Bush a stock market crash, a near-depression, and a paralyzed economy—and left to Trump a booming stock market, a recovered economy, and a low overall unemployment rate. But many millions of miners, factory workers, retail, service, and clerical workers remained unemployed or underemployed.[1] Trump's most appealing campaign pitch was that he (and he alone) could put America back to work again, bringing back the millions of jobs that had been outsourced to foreign countries. Globalization was an especially juicy target. Economists love it, multinational corporations love it, executives and shareholders love it, and so do consumers buying ridiculously cheap products. But for the millions of workers who have already lost their jobs, or fear losing their jobs, globalization is a hungry monster greedily stealing food from the table. Influenced by big campaign contributions and corporate lobbyists, leading politicians from both parties have consistently pushed free trade at the expense of the American worker.[2] The resulting outsourcing of jobs had been of huge benefit to big-money global players; a huge loss to the local little guys who are its collateral damage.

Trump's promise struck an understandably responsive chord among the dispossessed. His victory was sealed in the crucial midwestern Rust Belt states, precisely because he positioned himself as the people's champion, filling the vacuum left by both parties. Although Trump can deliver on his jobs promise in only the smallest, most purely symbolic ways, working people are understandably grateful to him for seeming to care.

Unfortunately, though, the underlying jobs problem has no quick or simple fix. Most of the jobs were lost to automation, not globalization, and sadly they will never return.[3] Our economy seems healthy, despite losing millions of jobs, only because technology has greatly increased productivity.[4] Computers started replacing people in lower-paying positions, but are now about to hit the professions. There are fewer and fewer workers safe from computer competition, and fewer and fewer skills that computers can't master. Our society is richer than ever, but the average person is in increasingly worse shape and prospects appear even bleaker for his children.[5]

Computers and robots could eventually eliminate almost half of our remaining jobs, with lost wages of $2 trillion a year.[6] They already do 10 percent of the work worldwide, but by 2025 they are projected to perform 25 percent.[7] Inevitably, despite Trump's empty reassurances, we will have far too many people chasing far too few jobs. In the old days, improved technology was a winner for almost everyone—increased productivity enhanced national wealth and provided workers with a higher standard of living. Jobs lost to the new technology were more than balanced by new jobs created. No more. For the first time in history, we are now experiencing the painful gap of productivity rising, while worker earnings stagnate. The faster technology races ahead, the further behind most people are falling. There have always been cycles of upheaval as technology displaced workers, but this cycle is uniquely ominous because machines are replacing people in every job category, except at the very top—leaving displaced workers no recourse to reemployment.[8]

Inequality

Our economy is the most vibrant in the world, but its benefits have not (as Reagan had promised) "trickled down" to everyone in anything approaching an equitable distribution. Technological advances always increase inequality, but the tilt in the United States has been ridiculously extreme—the richest 20 people in the United States now own more assets than the entire lower half of the population (that's 170 million people). Chief executives of major corporations used to earn 40 times more than their average employee; now they earn 400 times more. American corporations are sitting on record large cash reserves of $5 trillion, while their workers' wages continue to stagnate. And the richest people and richest corporations avoid paying their fair share of taxes because they also own the politicians (especially the Republicans), creating a rigged system that redistributes wealth from everyone else to the richest few. Perhaps the clearest and cruelest proof of inequality is that the life expectancy of middle-aged whites is declining because of alcoholism, opiate overdoses, and suicide. White, less-educated people are suffering greatly in a world of plenty. It is obscene that the Trump administration seeks to cut medical coverage for those who most need it to give whopping tax breaks to the very richest people who least need more money.

Trump and Bernie Sanders both gained tremendous traction mobilizing completely appropriate anger at Clinton for her close ties to superrich financial wheeler-dealers. Her husband had deregulated the banking industry in a way that encouraged wildly speculative, but very low-risk, gambling. "Too big to fail" banks had played a morally hazardous, "heads I win, tails you lose" game that left homeowners and taxpayers picking up the tab when the mortgage bubble exploded. George W. Bush and Federal Reserve chairman Alan Greenspan had been asleep at the wheel. And Bush and Obama exacerbated the pain by bailing out the bankers with public money, while doing little for the homeowners—the bankers laughed all the way to the bank; the homeowners cried all the way to foreclosure.

People had every right to mistrust Clinton's Wall Street connections, but trusting Trump was a classic jumping from a sizzling frying pan into a much worse fire.[9] Democrats have been far from perfect, but consistently promote more egalitarian policies—opposing tax giveaways to the rich, supporting programs for ordinary people, passing a higher minimum wage bill, and providing a safety net for the poor. Clinton's platform was economically much more favorable than Trump's to everyone except the corporations and superrich, but she lacked passion in defending the little guy and failed to display sufficient righteous indignation in exposing Trump's tax dodges, his proposed further tax giveaway to the superrich, and his serial cheating of the people who worked for him. Trump's cabinet of billionaires and billionaire wannabes nicely illustrates our country's inequality and will exacerbate it. His Treasury secretary embodies all that is ugly and unfair—a ruthless Wall Street sharpie who made one fortune promoting the mortgage bubble and then another by creating a mortgage foreclosure company that kicked people out of their homes. The straightest road to future Democratic victory will be to expose Trumpian hypocrisy and offer a meaningful alternative set of policies to right the great wrong of rapidly increasing inequality. The first, and most effective, step toward Democrats regaining the loyalty of working people would be for them to fight fiercely for a fair tax system and against Trump's proposal to tip the scales even further to favor the rich at the expense of everyone else.

Terror

People greatly exaggerate risks that are exotic and greatly underestimate much graver risks that are routine. Since 9/11, an average of only 9 people a year in the United States have died from terrorist acts; while each year more than 250,000 die from medical mistakes, 50,000 from drug overdoses, 37,000 from car accidents, and 33,000 from guns. But fear, rational or not, plays a huge part in how people vote. Trump has played the terror card crudely but expertly. He is the tough guy: "We're going to declare war on ISIS. . . . We will eradicate radical Islamic terrorism from the face of the earth." He provides no specific plan but proudly proclaims his intention to fight fire with fire—to be as brutal toward ISIS as

ISIS has been toward us—including torture (even beyond water-boarding) and killing the families of terrorists. Military personnel have been trained not to follow illegal orders to commit war crimes, but Trump says he won't take no for an answer: "They won't refuse. They're not going to refuse me. . . . If I say do it, they're going to do it."[10] Trump's tough-guy trash talk about terrorism was highly successful on the campaign trail, but is horribly dangerous coming from the White House.

Trump's position is un-American, violates international law, and offends common humanity. He also ignores the scientific evidence and military experience that torture simply doesn't work, and often backfires.[11] He has become an ISIS recruiter's dream—mobilizing Muslim hostility and brutality toward America in reaction to his obvious hostility and brutality toward Muslims. This kind of eye-for-eye vendetta mentality is the oldest and most primitive way of settling tribal disputes—revenge is psychologically sweet in the short run, but eventually self-destructive to both sides. As Martin Luther King put it, "If we do an eye for an eye and a tooth for a tooth, we will be a blind and toothless nation." I had a chance to witness firsthand the alienation of young Muslim men, working as an expert on the Boston Marathon bombing case. The Tsarnaev brothers were overly ambitious failures, not originally religious or politically active, who elevated their fairly petty personal grievances into a glamorous jihadi narrative. They might never have committed their bold terrorist acts had the United States treated terror as a dastardly crime rather than legitimizing it as an act of holy war. Trump didn't start the "war on terror" but he certainly is a genius at adding fuel to its fire. Conflating the criminals of ISIS with the billion peace-loving Muslims who detest terror as much as we do helps to create new Tsarnaevs. People who voted for Trump saw in him a powerful protector against the risk of terrorism, but my guess is, he'll be much more of a lightning rod for it.

Inexcusably and inexplicably, Trump constantly attacks the intelligence agencies meant to protect us from terrorism and even fails to attend most national security briefings. He trusts his gut and crazy conspiracy theories generated by right-wing media. Having competent, nonpolitical intelligence is crucial to counterterrorism. Instead, Trump has politicized and personalized national security and rendered it irrelevant and incompetent. Any real (versus reality TV) president would know that his very first priority is keeping tabs on the enemies, not fighting and castrating our intelligence services. There is also a great risk that when another terrorist attack occurs, Trump will use it as an excuse to assert emergency dictatorial powers that he may never relinquish.

Immigration

Trump capitalized on the natural human tendency to revert to tribalism during troubling and threatening times. Simmering prejudice can rise to a boil when there's realistic competition for jobs, cultural competition over language and values, and an inciting catalyst like Trump. Proximity can breed understanding and friendship, but it can also breed misunderstanding and prejudice. When I lived in the South in the 1970s and again in the early '90's, many people in rural areas had never met an immigrant. The towns had no Mexican, Indian, or Chinese restaurants. Now immigrants are everywhere, including in rural America, often doing jobs that no one else is willing to do.

Trump was skillful in transforming free-floating anxiety about a declining America into focused prejudice against the immigrants who are supposedly responsible. "The U.S. has become a dumping ground for everybody else's problems. . . . When Mexico sends its people, they're not sending their best. . . . They're sending people that have lots of problems, and they're bringing those problems with us. They're bringing drugs. They're bringing crime. They're rapists. And some, I assume, are good people."[12] We are good; almost all of them are bad. This is our land, this is not their land. It's comforting to concentrate crime fears by imagining that all criminals are foreign. It's comforting to imagine that job insecurity and low salaries can be solved simply by getting rid of immigrants. It's comforting to reduce complex problems to us-against-them solutions. Trump is a master at stirring up nationalist passions, xenophobic fear, vitriolic anger, and misplaced righteous indignation. Keeping migrants out of the country became his signature issue. It hit all his bases: fears about jobs, terror, and crime; hatred of globalization; uneasiness about cultural change; anger at unfairness; aspirations to keep Americans first; helping the disrespected feel heard and heeded. The immigration "problem" also seemed oh so simple to solve. Kick out the undocumented. Build a high and beautiful wall on the Mexican border to keep out all those "rapists," "drug dealers," and robbers of American jobs (and make Mexico pay for it). Block Muslims from getting into the country and closely surveil those already here.

Some of Trump's immigration policies were unconstitutional, many were terrible for the economy, most were impractical, and all were heartless, unfair, and unnecessary. Everyone living in the United States descends from immigrant ancestors. Every immigrant wave, once settled and established, has welcomed the next with much less than full enthusiasm. The greenhorn might provide necessary skills, but he or she also arouses understandable fears for lost jobs, lost customs, lost prerogatives, and lost familiarity. Trump, a serial breaker of promises, felt compelled to keep his ill-considered promises on immigration. Within his first week in office he issued executive orders to build the wall and to block migration from seven Muslim countries. Both decisions were made quickly, impulsively, incompetently, and without the usual consultations to make sure they are practical and legal. Both were unnecessary and both had dire consequences (possible trade war with Mexico, widespread demonstrations, and a constitutional crisis over religiously discriminatory visa restrictions).

The reality is that the United States has always needed immigrants to function properly, and it still does. They provide a large net economic plus to our economy by filling labor shortages in construction, agriculture, service industries, science, medicine, and technology. Up until Trump, America has attracted the best and the brightest from all over the world. Of the 579 Nobel Prizes ever awarded, 353 were won by Americans—almost one-third of whom were immigrants. The trend toward imported brilliance has recently strengthened—all 6 of last year's U.S. Nobel Prize winners were foreign-born. And Silicon Valley wouldn't be a world tech leader if it couldn't draw its workforce from all around the world. Trump's xenophobia is scaring away people we need and will encourage a brain drain of those already here. We must contain immigration to safe and manageable levels, but we can't live without it.

Government Not Delivering

"In this present crisis, government is not the solution to our problem; government is the problem"; "Man is not free unless government is limited"; "The most terrifying words in the English language are: I'm from the government and I'm here to help"; "Government's view of the economy could be summed up in a few short phrases: If it moves, tax it. If it keeps moving, regulate it. And if it stops moving, subsidize it." All the above quotes are from Ronald Reagan, who began the clever Republican strategy of running against the government he was simultaneously leading. They helped persuade voters to elect Reagan, but the strategy turned out to be a scam. Under Reagan's crony capitalism, the number of workers in the federal government actually expanded by 324,000 people and the federal deficit tripled from $907 billion to $2.6 trillion.[13] Republican presidents have copied the paradoxically successful pattern of blaming everything on the government, while they run it with egregious incompetence. Tearing down the government was also a strategy for allowing the giant corporations to win lucrative "outsourced" contracts and run free of the regulations necessary to protect workers, consumers, and the environment. Propaganda blinds people to the benefits of government (Social Security, Medicare, free education, police, infrastructure, etc.) and focuses attention instead on its shortcomings, some of which are self-inflicted, by the very politicians doing the blaming.

Trump ran as the outsider, populist champion of the little guy fed up with the corruption of the political elites. He promised to "drain the swamp." Instead, he has created an unprecedented new swamp—using the presidency to further his family's business interests. He, his children, and in-laws shamelessly use his position to peddle Trump products—condos, golf memberships, hotels, clothing, perfumes, and jewelry. The hypocrisy of his immigration stand is baldly exposed by his family's willingness to sell visas in return for a $500,000 real estate investment. There have been presidential scandals of all sorts in the past, but nothing approaching the absence of ethics in the Trump White House.

Trump's "deconstruction of the administrative state" is a recipe for disaster. He has appointed a cabinet filled with billionaires, doctrinaires, conspiracy theorists, incompetents, unprepared political hacks, special interest apologists, fools, and knaves. It is perfectly reasonable for people to distrust and blame our government for not doing a better job, but it was perfectly unreasonable to turn to Trump for a solution. His business career was characterized by disorganization and foolish risk taking that led to five bankruptcies. His presidency is chaotic, creating self-inflicted crises on an almost daily basis. The job of his appointees is to dissolve the government, and their incompetence and insincerity are prime qualifications for accomplishing it. In "Trump world," all the power will reside with the president and the corporate executives, unopposed by the will of the people or the power of the press (now known as the "opposition party"). Trump is undertaking this radical revolution despite the fact that he captured only a minority of the popular vote, won because of Russian rigging, has the worst poll ratings in history, and faces massive public opposition. You can't make stuff like this up. Trump isn't crazy, but the Trump White House certainly is.

Rapid Change

The world seems to be spinning ever faster. When my father was born in 1900, there were no cars, electric lights, or telephones. He remained forever amazed by air travel, TV, trips to the moon. When I was born in 1942, there were no computers, smartphones, Internet, decoded genes—and I am just as amazed as he was by the rapid changes wrought in just seventy short years. Technological revolutions always privilege the young, indigenous users at the expense of older cranky traditionalists. My grandson looks forward eagerly to exponential leaps in genetic engineering and artificial intelligence—both scare me to death as a threat to our basic humanity.[14] Many (especially elderly) voters picked Trump in the belief that he was a conservative who would help preserve the past, stabilize the present, and reduce the risks of frightening changes in the future.[15] He promised to return America to past glory days when it was the dominant world power; white men ruled and women and minorities knew their place; where gays stayed in the closet and people kept the sex they were born with; when the concept of "political correctness" didn't exist; and when it was OK to mine coal without worrying about ruining the environment for our children.

Anyone who expected Trump to be a "conservative" preserver of the past was badly misled. Trump is, without any close equal, by far the most radical president in United States history—making drastic changes now and opening the door to even more terrifying changes in the future. Real conservatives, by temperament and experience, understand that big and sudden changes always carry harmful unintended consequences—things should be fixed incrementally and only if broken. Radicals push blindly for quantum leaps, heedless of consequences. Trump would have been right at home during the Reign of Terror in the French Revolution. He is a radical wolf hiding in conservative sheep's clothing. By comparison, Obama and Clinton are cautious conservatives and temperamental moderates.

The Attitudes Trump Is Exploiting

Follow the Leader

Theodor Adorno, a victim and sophisticated observer of Nazi Germany, used psychology as one way of understanding why people had so readily succumbed to its fascist takeover. A survey he conducted in the United States revealed that many Americans also have the characteristics of what he called "the Authoritarian Personality." These include strongly defending conventions; being submissive to those above, and domineering to those below; devaluing intellectual activity; overvaluing power and toughness; blaming others; being cynical; and believing conspiracy theories and superstitions. People with this "Authoritarian Personality" obey, rally to, and sometimes become powerful and dominating leaders. And they respond aggressively to outsiders, especially when they feel threatened. By acting tough, Trump displays his own (and plays to his followers') authoritarian inclinations.

Adorno, writing in 1950, predicted Trump's ability today to peddle his "alternative reality": "Lies have long legs: they are ahead of their time. The conversion of all questions of truth into questions of power not only suppresses truth, as in earlier despotic orders, but has attacked the very heart of the distinction between true and false." And Adorno also predicted the mentality of a vocal minority of Trump's followers: "There was no difficulty in finding subjects whose outlook was such as to indicate that they would readily accept fascism if it should become a strong or respectable social movement." Adorno feared that fascism could become respectable through TV, radio, and film propaganda. Remember that this was written a half century before the reality TV show that anointed Trump, the talk-radio conspiracy peddlers that promoted him, and the Twitter that spread his twisted gospel. Adorno could see in the appeal to Americans of 1950s McCarthyism the seeds of twenty-first-century Trumpism.[16]

A close study of Adorno informed the polling results of Matthew MacWilliams, one of the few prognosticators who gave Trump a serious shot at winning the 2016 election. He succeeded, where others failed, by including four Adorno-like items in his survey of 1,800 likely voters. It turned out that these questions (for example, whether you raise your kid to be more respectful or more independent) were the best predictors of support for Trump. On the eve of the election MacWilliams wrote: "Trump support is firmly rooted in American authoritarianism and, once awakened, it is a force to be reckoned with." The force of Trump's authoritarianism is now fully awake, and it surely must be reckoned with.[17]

Overconfidence in a Con Man

"It ain't what you don't know that gets you into trouble, it's what you know for sure that just ain't so." Trump says our world is broken and that he (and again he alone) can fix it. His policy proposals don't remotely add up—there is no conceivable way to simultaneously pass a massive tax cut, fund huge increases in military and infrastructure spending, and reduce the national budget deficit. But Trump promises all three and his followers blindly believe him. Trump lies with apparent pleasure and aplomb. The media can barely keep up with all the fact checking necessary to refute all of Trump's constant "Pinocchios"—most of which are childishly obvious and patently false. But the transparency of Trump's deceptions does not discourage his faithful followers from accepting that he is truthful and that the reporters he hates are the "most dishonest people on earth."

In a fearful and uncertain world, Trump is ever the clever confidence man, cynically trading on the overconfidence that is an inherent part of human psychology. He embodies within himself, and unconsciously exploits in others, the "Dunning-Kruger effect." Experiments done by these two Cornell psychologists, and others, show that the people with less ability at any given task are more likely to overestimate their own skill and underestimate the skill of others. In contrast, more competent people tend to underestimate their own competence and assume that, because they can do something easily, everyone else can also do it just as easily. The effect holds up across a wide variety of activities: academic tasks, sports, chess, driving, and the practice of medicine.[18] Dunning and Kruger put into quantitative terms what has been the wisdom of the ages. Those who are ignorant in a certain area often don't know enough to know what they don't know. People who know a lot usually know how little they know. If you don't know what you don't know, you can't correct your ignorance. If you don't know when you are making a mistake, you will keep making it. As Shakespeare put it: "The fool doth think he is wise, but the wise man knows himself to be a fool."

I believe this explains a lot about Trump—a man of overweening overconfidence, unchecked by knowledge or wisdom, and unable to learn from his mistakes. It also illuminates why Trump voters were so willing to forgive his mistakes and fiercely attack those who point them out. And finally, it explains why the mostly better-educated Clinton voters were so surprised that the mostly less-educated Trump voters had so much trouble seeing through him. Trump successfully stoked the electorates' fears and sold himself as the magician who could make them disappear. Contradictory facts and expert opinion were irrelevant—many were ready to accept Trump's gut feelings as the best guide to our nation's policy. The world would have been a much safer place if Trump were much less sure of himself and his followers were much less sure of him.

Misogyny

By the end of the 2016 campaign, it became clear that many people in the United States, women as well as men, were simply not ready for a woman president. The simply awful pins on display at the Republican National Convention said volumes about how misogyny factored into the 2016 election: "Don't be a pussy. Vote for Trump in 2016."; "Trump 2016: Finally, someone with

balls."; "Trump that bitch."; "Hillary sucks but not like Monica.";
"Life's a bitch: don't vote for one."; and "KFC Hillary Special. 2
fat thighs. 2 small breasts, left wing."[19] General misogyny, and
specific distrust of women in politics, runs both long and deep in
American culture. Recall that our Declaration of Independence
affirmed that "all men are created equal" but was stunningly si-
lent on female equality. Thirteen years later, our constitution pro-
vided voting rules that continued to disenfranchise women, who
did not finally get the vote for 130 more years until 1920 (a full
fifty years after black men and only after the bitterest of battles).
The first proposal to add a women's Equal Rights Amendment
to the Constitution was drafted in 1923, but it then had to be
reintroduced in every session of Congress, until finally it passed
in 1972. Jubilation grew when thirty of the thirty-eight required
states approved it within just one year. But momentum mounted
against equal rights for women, spearheaded mostly by conser-
vative women's groups, finally killed the amendment and it now
appears dead forever.

The United States is far behind the world in electing a female
head of government—seventy-nine women have already achieved
this distinction in other countries[20] and many countries have also
approached gender equality in their legislatures. Hillary Clinton
was in a no-win bind—too masculine a woman for some voters,
not enough of a man for others. This paradox may continue to
confound succeeding female contenders to what has become an
exclusively male club. It is a peculiarly female burden to be simul-
taneously resented for not being cuddly enough and disparaged
for not being strong enough. Many people supported Trump be-
cause they saw him as a strong Big Daddy (or Big Brother), able
to protect and fight for them. It's unclear whether our fear of a
Big Mommy is more deeply psychological or more superficially
sociological, but either way, it still dominates American poli-
tics. The radical right-wing attack dogs are already turning their
defamation machine on Elizabeth Warren, a woman unafraid to
challenge sacrosanct male hegemony.

Racism

Donald Trump's racism is long-standing and unusually well documented. At the very beginning of his career in 1973, he began a bitter two-year fight with the U.S. Department of Justice for refusing to rent apartments to African-Americans. And more recently, he wasn't shy about showing he hasn't changed much: "Black guys counting my money! I hate it. The only kind of people I want counting my money are short guys wearing yarmulkes. . . . Laziness is a trait of blacks."[21] White supremacists, Klansmen, militiamen, neo-Nazis, and other extremist hate groups were thrilled to have their usually roundly condemned prejudices brought into the mainstream and legitimized by a United States president. But it also struck a responsive chord among many otherwise quite decent white Americans, who feel threatened by and unhappy with the rapidly decreasing white predominance in an ever more diversified United States. During the first half of the twentieth century, the United States was almost 90 percent white; today whites account for 63 percent of the population (and falling), a remarkable demographic shift. Each year, more whites die than are born and there are more births of nonwhite than of white babies. By the mid-twenty-first century, whites will be a minority in what they have heretofore regarded as their backyard. A black president had been a hard swallow and final straw for some—giving Trump the cheap opportunity to exploit a defiant white backlash vote. "Make America Great Again" was a thinly veiled code for making America white again.

Social Conservatism

It is surprising that Trump won the almost unanimous support of evangelical voters, given the fact that few people in the history of the planet have so consistently and pervasively violated the basic Christian values of sexual continence, charity, humility, fair play, forgiveness, truthfulness, and high ethical standards. Trump is a serial sexual predator, a bombastic self-promoter, an unabashedly vindictive injustice collector, an extravagant liar, and a chronic business and tax cheat. Not exactly the recommended preparation for the kingdom of heaven. And before switching cynically to fool conservative voters, Trump had held typically New York City libertarian views on abortion and gay rights. So add hypocrisy to his sins. None of Trump's extensive catalog of immoralities discouraged most evangelical leaders from enthusiastically jumping on his bandwagon. They constitute the purest political opportunists of our time—willing to ride any horse, however disreputable, if it will further their personal power interests.

Libertarianism

Trump's victory can be attributed in some measure to libertarian votes—both those that went directly to him and those that might have gone to Clinton had there not been a Libertarian Party alternative. I greatly treasure and exercise my personal liberty, but I am not a blind, card-carrying libertarian. Your freedom must end when it begins to threaten my liberty or my life. And often, government regulation is the only way to ensure that you are not infringing on my rights when you are exercising your own. You can't play baseball without having rules of the game that establish a level playing field, or umpires empowered to enforce them. Many voters picked Trump because they trusted him to reduce government's role in the running of our country and their lives. They resent government regulations on business, guns, land management, and protection of civil rights.

I disagree and strongly support the role of government in all these activities, not because I trust it so much more than do the libertarians, but rather because I trust the goodwill and good sense of corporations and individuals so much less. Without business regulation, greedy corporations exploit the public (think drug prices, mortgage crisis, pollution, global warming, worker safety). Without gun control, we have one of the world's highest rates of homicide and unnecessarily high rates of suicide. Without land management, we have no protection for national parks and wilderness, and public lands are exploited by the privileged few, not preserved for the public many. Without federal enforcement of civil rights, we have unconscionably discriminated against minorities, gays, and women. Too much government stifles people and the economy; too little leads to inequity in the present and the neglect of our obligation to the future. The ultimate irony, of course, is that there has never been a bigger threat to American liberty than President Trump's dictatorial instincts.

Attitude Toward Truth

Aeschylus, the great tragic playwright of ancient Greece, made the still timely observation that "in war, the first casualty is truth." In the no-holds-barred U.S. political wars, bold untruth has become the most powerful of all political weapons. Ultraright-wing talk radio, conspiracy theory Internet sites, and Fox News spew forth a constant spate of alternative facts and extreme opinions that are often outright lies and always anything but "fair and balanced." They follow the chilling advice of Hitler's propaganda mentor, Joseph Goebbels: "It would not be impossible to prove, with sufficient repetition and a psychological understanding of the people concerned, that a square is in fact a circle. They are mere words, and words can be molded until they clothe ideas and disguise."

Trump understood that people who feel desperate, anxious, angry, and helpless are not in a mood to listen to rational arguments. His fearmongering pitch is that we are now living in the worst of worlds, in the worst of times; that there are even worse dangers ahead; that enemies lurk on all sides; and that we can trust him to keep us safe. He daily succeeds in passing off a fusillade of "alternative facts" (aka bare-faced lies) because frightened people are ready to accept them. Human irrationality in the face of stress has a long past and may, unfortunately, also enjoy a great future.

The two sacred priorities for honest media (Fox News personnel can stop reading now) are truth and neutrality. Trump made these incompatible. His nonstop lying has forced the press to abandon its neutrality in order to report the truth. The most dishonest man on the political stage is constantly attacking the press for being dishonest. He stubbornly defends outright lies by accusing the press of falsehoods. Trump had to declare this all-out war on the mainstream media because it is the last major threat to his autocratic ambitions. Contrast Trump's view on the press with Thomas Jefferson's: "Our liberty cannot be guarded but by the freedom of the press, nor that be limited without danger of losing it." American democracy has never been so dependent on truth telling and fact checking by the much derided "mainstream" media.

Trouble is, many voters no longer get their news from newspapers and instead are much more influenced by tweets, radio rants, conspiratorial websites, and un-fact-checked social media. In our hyperwired world, it is more true than ever that "lies can travel around the world in the time it takes the truth to put on its shoes." All but 8 of the nation's 208 newspapers endorsed Clinton—because they trusted her competence and basic decency and were terrified by Trump on both counts. This overwhelming vote of no confidence in Trump was cast even by all but a handful of the nation's Republican papers, some of which had never in their entire history ever backed a Democrat. But Trump won and Clinton lost because tweet wars had become far more important than serious reporting—and hothead mouthing off by conservative radio hosts more influential than carefully considered editorial endorsements.

Trump not only fights the media, he cleverly exploits it. Prior to his presidential run, Trump was widely regarded as a vainglorious clown. Trump hasn't changed a bit—it was the crowds' reaction to him that changed. I don't get his entertainment value, but to his adoring fans, Trump's stream-of-consciousness stump speech was as funny as stand-up comedy, as uplifting as a church tent revival, and as cathartic as a professional wrestling match. An EEG study of thirty people on both sides of the political spectrum, watching the Trump and Clinton debates, found that Trump held their attention significantly longer and kept them in a higher state of mental arousal, whether or not they agreed with him. He is an entertaining and perversely captivating performer.[22]

Trump obviously loved all the attention, easily turning the campaign into a raucous reality TV show—a turf on which he is an accomplished performer and his opponents were sixteen overwhelmed straight men and two mostly frozen straight women. The media is now doing its best to keep dishonest Trump honest, but it had built him up during the campaign. In the never-ending quest for higher ratings, it gifted him the equivalent of $5 billion in free airtime exposure.[23] Political campaigns are usually sober, dull affairs. But for those who loved him, Trump's daily antics and baroque lies made politics fun again. Andy Warhol and Marshall McLuhan would not be surprised. Rome pandered with free bread and the entertainment value of the circus. We have Trump.

America First and Great

I was born during World War II, at the zenith of U.S. predominance on the world stage. Growing up, I was justly proud that the United States had the most advanced technology, the bulk of Nobel Prize winners, the best movies, the most political freedom, the greatest athletes. It felt like we led the world in both greatness and in goodness. It is now getting more and more difficult to make any of these claims, to feel an unvarnished pride in ourselves, and to fend off criticism from foreign friends.

Many patriotic Americans (especially the older ones, who have lived through and feel keenly our rapid fall from grace) have responded emotionally and gratefully to Trump's grandiose promise to "make America great again." And lots of people have wondered why we have wasted billions on foreign wars and infrastructure programs, while our own infrastructure is so poorly maintained; why we support aid programs for the needy in other countries, while our own needy are neglected; and why other countries (think China) seem to outsmart us in trade deals. Perhaps Trump's proven skill at getting the best deal for himself would translate into his getting a much better deal for our country. They were encouraged by his inauguration speech, warning other countries: "From this day forward, it's going to be only America first, America first." Trump voters, aggrieved at the tarnishing of the American dream and by their lack of personal reward for loving their country, looked to him as the country's savior. Trump is great, so he can make America great; he will protect us from our external enemies and root out our enemies within. This is dangerously demagogic stuff in a democracy, but it is appealing to many who are disgusted by America's humiliations and by the sloppy ineffectualness of legislative gridlock.

Trump may be a knight in shining armor to those looking for bold, simple solutions—but he is a two-bit, would-be Mussolini to those fearing his disrespect for democracy. And Trump's "America First" slogan carries the heavy baggage of a noxious, xenophobic, and racist past. The Native American Party (appropriately nicknamed "Know Nothings"), composed exclusively of white Protestant men, was formed in the 1850s to block the migration of German and Irish Catholics. They claimed that loyalty to the pope in Rome precluded Catholics from ever being trustworthy citizens of the United States.[24] Sounds a bit like Trump today on Muslims. The America First Committee was a more recent, even more troubling, manifestation of the same insular attitude. Formed to keep us out of World War II, its most prominent spokesman was the heroic aviator Charles Lindbergh—who also happened to be anti-Semitic and an admirer of Hitler and Nazi Germany.[25] I'm reminded of Trump's bizarre "bromance" with Russian president Vladimir Putin. The modern essence of America First is American military and trade unilateralism—a go-it-alone independence, free of entangling alliances or a sense of responsibility to adhere to international treaties and norms. This may have been a perfectly reasonable stance for George Washington's young republic, but it is impossibly self-destructive in our tightly intertwined, deeply interdependent world. Trump fancies himself a man on horseback rescuing a diminished America, but he is very much more a blind and clumsy bull in a china shop, trampling on its most precious values and institutions.

The Triumph of Fake Populism

One man's populism is another's demagoguery. Real populism should be the stuff of everyday good government—the effort to ensure the rights of the everyman, protecting him from the greed of the powerful elite. Fake populism is the seduction of the masses by demagogues who promise everything before gaining power, but afterward deliver nothing but exploitation. The word *demagogue* is from the ancient Greek "leader of the people"—a seemingly benign enough meaning, but quickly gaining pejorative connotations through painful experiences with not-so-great leaders. There is nothing new under the sun—demagogues are similar in all times and all places, as are the conditions that spawn them. Wherever and whenever there are democracies, there will also be demagogues. And they will all use emotion, oratory, and impossible promises to exploit the people for their own selfish ends. Aristotle's description of the Athenian demagogue Cleon, 2,400 years ago, nicely captures Trump today: "He was the first who shouted on the public platform, who used abusive language, while all the others used to speak in proper dress and manner."

Populism has historically been the tool of autocrats and the graveyard of democracy. Athens had to institute an ingenious system of ostracism to protect its people and democracy against potential tyrants. The citizens could impose an honorable ten-year exile for anyone they feared would become too powerful. No one was immune—Themistocles, savior of all Greece during the perilous Persian Wars, its wisest and most successful leader, was nonetheless ostracized as a threat to democracy soon after the war had been won. Athens understood that democracy is fragile and easily corrupted. The people must be protected from the powerful and from their own instincts to follow them. In Aristotle's words, "Revolutions in democracies are generally caused by the intemperance of demagogues."

Demagogues may be phony, but the problems they exploit are all too real, and common. Populist movements arise out of a sense of shared grievance among ordinary people against a government they experience as disinterested or hostile to their needs. The crises giving rise to populist movements are similar everywhere, but the proposed solutions dichotomize. Right-wing insurgencies, like Trump's, promise a return to a halcyon era that never existed. A reviled "other" is scapegoated as the impediment to achieving order, prosperity, and respect. Left-wing insurgencies unite the 99.9 percenters against the 0.1 percenters, with promises of reducing wealth inequality. Right- and left-wing demagogues look very different before assuming power, exactly alike after they become dictators. Trump is in the early stages of executing strategies made familiar by recent demagogues like Stalin, Hitler, Mussolini, Mao, Franco, Pinochet, Peron, Idi Amin, and a whole bunch of tin-pot third-world autocrats. He differs from them, not in intention, but in ability and plausibility—falling among the least competent of history's would-be tyrants. Trump got lucky because his message fell on well-fertilized soil and his reality TV celebrity gave him a ready-made fan base.

Radical right-wing populism has always been more top down than bottom up—more AstroTurf than real grassroots. The John Birch Society was started in 1958 by twelve rich guys, including Fred Koch, father of radical right-wing patrons Charles and David Koch. Its platform was so kookily extreme that William F. Buckley Jr. (himself the founder of a pretty extreme form of U.S. conservatism) denounced it as "far removed from common sense" and fought any role it might have in the Republican Party. He eventually lost and the Birch worldview won because he had only clever words at his command, while they had very big money.[26] Today's Republican platform, prejudices, and policies are derived almost plank for plank from the Bircher doctrine. The Koch brothers have been the most influential moving forces in turning extremist doctrine into mainstream Republican policy—and selling it to the common people it helps to fleece. They (and their buddy billionaires) have spent tens of billions of dollars creating fake grassroots organizations, political think tanks, an army of political operatives at the state and local level, and training camps for conservative lawyers and judges.[27] The enormous efforts promote science denial, tax breaks for the wealthy, deregulation, pollution, global warming, and minority bashing. Unholy alliances have been formed with the tobacco industry, the National Rifle Association, and extremist religious leaders. Fake populism's biggest success story is the Koch-conceived, Koch-funded Tea Party—which first conquered the Republican Party, then seized the White House.

The top of the heap is expert at perverting populist ideology for their own, cynical and sinister, elitist ends—protecting their power and privilege by playing the "divide to conquer" game. Brilliant political propaganda skillfully co-opts the underclass it is screwing. The legitimate grievances of poor whites, who receive an ever-shrinking slice of the American economic pie, are redirected against blacks, Latinos, women, and immigrants. The elites keep their rich spoils (and tax loopholes) by stoking inchoate fears and tribal feuds, and offering trickle-down crumbs.[28] Attacks on "big government" protect the elite from the one institution that might umpire a fairer distribution of wealth. Radical right-wing demagoguery feeds upon and promotes all our societal delusions—using them as disguise for robbing the public purse.

Conspiracy-Theorist-in-Chief

Conspiracy theory, as a tool in our politics, is as old as our country and as American as apple pie. In uncertain times, scared people seek, and hold stubbornly to, a comforting certainty—especially when confronted by discomforting truth. Conspiracy theories evolve out of a set of interacting false assumptions: everything is purposeful, nothing random; everything connects to everything else; appearances all deceive; people are malevolent; and someone is to blame. Conspiracy theory provides an explanation, an excuse, a villain, and a call to action. It is supported by cherry-picked evidence and is fairly immune to disproval by reality.

When I was at Columbia College, in the early 1960s, a professor named Richard Hofstadter wrote a seminal book on *The Paranoid Style in American Politics*. He was describing the lunatic fringe, radical right—then a small voice met with bemusement by liberals and with strong opposition by traditional conservative Republicans. Hofstadter wouldn't be surprised, but would be dismayed, that fifty years and billions of propaganda dollars later, the radical-right conspiratorial worldview now dominates the White House and Congress. "The idea of the paranoid style as a force in politics would have little contemporary relevance or historical value if it were applied only to men with profoundly disturbed minds. It is the use of paranoid modes of expression by more or less normal people that makes the phenomenon significant."

The Koch brothers are probably more or less normal people, but grew up in a strict John Birch household that nurtured in them some pretty strange beliefs. With great effort, energy, money, and skill, they have now transformed their fringe radical-right party into the politically dominant force of our time. It is a paradoxical tribute to their accomplishment that the greats of Republican Party history—Abraham Lincoln and Teddy Roosevelt—could not today survive a Republican primary because of Tea Party opposition. More recent Republican presidents would also not pass Koch muster: Eisenhower was called a communist by Papa Koch; Nixon passed progressive climate and social programs and engaged China; and even Reagan was perhaps too moderate for today's "Republicans." The Kochs despise Trump as a grotesque Johnny-come-lately who crashed their party, but they have succeeded in co-opting his administration with their own carefully selected henchmen.

Those determined to believe Trump is crazy miss this context. They find confirmation in the fact that he says and tweets crazy things—and then doubles down on them even when they are proven to be patently false and without any credible supporting evidence. Within his first months in office, Trump stuck to two bizarre claims of election rigging against him—that Clinton voter fraud robbed him of five million votes and that Obama is a sick/ bad person who wiretapped his office. Trump's psychological and political motivations are all too transparent. His false conspiracy theories of election rigging against him counteract the truth that Putin had in fact skillfully rigged the election for him (likely with the connivance of Trump and cronies).

Projection is the psychological defense of attributing to your enemy your own deeds, thoughts, and impulses. Conspiracy theory is the political defense of obscuring embarrassing revelations within a haze of countervailing accusation. We don't know whether Trump actually believes the weird stuff he says or whether he is just using it as part of his con game with the American people—and it doesn't really much matter because the results are essentially the same. But I suspect it is both. When in trouble, Trump searches for conspiracy excuses, swallows them whole, and then vomits them out. This is dangerous and despicable, but not crazy and delusional. A delusion is a fixed, false, bizarre belief that is exclusive to you and causes impairment. If 46 percent of the voting population joins you in the weird belief and elects you president, it is society's delusion, not yours. The only hope against well-publicized, well-funded conspiracy theory is a vocal, unintimidated free press— precisely why the media is Trump's "opposition party."

Political Polarization

On leaving office as our first president, a fearful George Washington solemnly warned the fledgling republic that political polarization was a serious risk to its uncertain future survival: "It agitates the community with ill-founded jealousies and false alarms, kindles the animosity of one part against another, foments occasional riot and insurrection. It opens the door to foreign influence and corruption, which finds a facilitated access to the government itself through the channels of party passions." Sounds like Trump's

America. We have always had our fair share of politically polarizing demagogues—we just never before elected one president.

My father was the ultimate political cynic. He would vote against the incumbent in every election, regardless of party affiliation, on the distrustful assumption that length of tenure in office allowed politicians to perfect the skills and networks that facilitate corruption. Any given newcomer was likely to be equally (or even more) corrupt than the old guy, but it would take him some time to catch on to all the tricks of the trade. My level of cynicism has never quite matched my father's, but it was sufficient until now to keep me politically inactive and unwilling to attach any great hope to either party. This was perhaps justified because, until the 1960s, there was so much overlap between the two parties. Temporary shifts in electoral success mattered some, but did not result in seismic shifts in policies. No longer is this the case and no longer is there any justification for neglecting politics. The differences separating the political parties are now clear, consistent, dramatic, and seemingly beyond compromise. Elections have become high-stakes gambles, with U.S. democracy and world sustainability both in the pot. If we don't cure our political malaise, it will kill us.

"Party sorting" has dominated and polarized American politics of the past fifty years.[29] It started when a southern Democrat, President Lyndon Johnson of Texas, aggressively pushed through the Civil Rights Act of 1964—against the fierce and stubborn opposition of the southern Democrats in Congress. Ever since the Civil War, the South had been solidly Democratic (never forgiving Lincoln's Republican Party for that war and for ending slavery). The South had also always been solidly conservative in its social, economic, racial, religious, and military values. The Republican "Southern Strategy" (used first in Barry Goldwater's bid for the presidency in 1964 and perfected by Nixon in his successful 1968 and 1972 presidential runs) succeeded in suddenly turning the solidly Democratic South into the reliably Republican South. This made the Republican Party much more homogeneously conservative and the Democratic Party more liberal—with little overlap.[30]

We now have the greatest degree of polarization between parties since the post–Civil War period of Reconstruction.[31] The ever-widening gulf came from the Republicans moving to the extreme right, while Democrats remained pretty much where they were. Hillary Clinton, today, is equivalent to a typical moderate Republican of fifty years ago, while most Republican politicians today would have been quite comfortable joining the radically extremist John Birch Society, which once seemed to be so kookily conspiratorial. Eisenhower and Nixon were much closer to today's Democrats than to today's Republicans. Even Reagan and the Bushes qualify as solid moderates compared to the current GOP radicals. And Clinton would be a dead-center politician in most European countries, while Trump and the current GOP would be more radical and right-wing than even their radical right.

With progressive political polarization has come increasing partisan hatred. A 2014 Pew poll of 10,000 adults showed increasingly strong antipathy against the opposing party (among Republicans up to 43 percent, from 17 percent; among Democrats up to 38 percent from 16 percent) and fear it is threatening the nation's stability (70 percent among Dems and 62 percent of Republicans).[32] Those, on both sides, with more extreme partisan positions participate more fully in the political process, driving their parties further apart than makes sense to the more moderate people in the middle. And the moderates are disappearing (39 percent, down from 49 percent). Politically interested people also flock only with political birds of the same feather (63 percent of Republicans and 49 percent of Democrats).[33] The scariest finding was that more than half of Americans expressed dissatisfaction "with the way democracy is working."[34]

Which brings us to Trump. His presidential success would have been inconceivable but for our current political polarization and political hatred. Trump's egregious incompetence, heavy moral baggage, and self-absorption would have made his election impossible but for his unique competence in stirring partisan hatred and exploiting political polarization. Though elected by a minority of the population, President Trump is making dictatorial moves, confident his base will follow because our political polarization gives them no alternative middle course. Political sorting has created an inviting vacuum this would-be dictator would love to fill.

All of Our Elections Are Rigged

When Trump was a nine-to-one underdog, fearful of losing in a giant landslide to Hillary Clinton, he complained constantly that the election was rigged against him. It turned out that the election was indeed very heavily rigged, but all in Trump's favor—and just enough for him to eke out a narrow Electoral College victory despite losing the popular election by almost three million votes. Typical of Trump, he then invented (and keeps doubling down on) the outlandish lie that he would have won the popular vote by millions if not for massive (and completely imaginary) voter fraud. Clinton's once seemingly unstoppable campaign momentum had been surprisingly derailed at the very end when three of her old enemies tinkered with our democratic voting process: Vladimir Putin on behalf of Russia, James Comey on behalf of the FBI, and Julian Assange on behalf of himself.

But the rigging of our political system goes far beyond the particularities of this one election and is deeply built into all our electoral processes. Democratic candidates have won the popular vote in every presidential election but one since 1988 (George Bush was the exception in 2004). Democrats also usually outpoll Republicans in the total national popular vote for both houses of Congress. Despite this Democratic preponderance of public support, Republicans now dominate the political process more powerfully than they have since the days of Reconstruction, just after the Civil War. They have firm control of all three branches of the federal government and have won two-thirds of governorships and two-thirds of state legislatures. And, remarkably, the GOP has attained this unprecedented level of power while pushing a radical platform that has repeatedly failed to deliver on its promises and is markedly out of tune with what most Americans want and need. This disconnect between power and popularity occurs because our political system is heavily stacked against the Democrats and heavily stacked for the Republicans.

The Founding Fathers were hardheaded, no-holds-barred politicians who constantly engaged in political insult and factional strife—anything but the saintly innocents portrayed in high school history books. They lacked the French Revolution's infatuation with the perfectibility of man and instead regarded us as selfish, desiring creatures, tamed throughout most of history only through the constraints that had been imposed by autocratic power. The United States could escape this past only by establishing strong republican and democratic institutions, using a comprehensive system of checks and balances to temper and defuse our baser natures. Their goals were to avoid both the tyranny of the many and the tyranny of the few. It was not easy to gain approval for a closer union from the then very separate, and often quarreling, states. The Constitution of the United States, now a sacred document, originated in a series of hotly contested (and narrowly won) compromises, meant to protect the rights of the smaller states and of the propertied few against domination by the bigger states, the great unwashed, women, and slaves.

The Senate apportions two seats to each state, large or small—a concession crucial to winning the support of the smaller states for ratification of the Constitution. This system was not so disproportionate an allocation of political power when all the states were relatively small and less unequal in population, but it has since become badly unbalanced. Each California senator (now always Democrat) today represents almost twenty million people; each Wyoming senator (now always Republican) represents fewer than three hundred thousand. So, each voter of Republican Wyoming has sixty times greater Senate influence than each voter in Democratic California.

The Electoral College system creates a similar bias in favor of smaller states. Hillary Clinton lost the Electoral College, despite her large victory in the popular vote, partly because Democratic voters are so heavily concentrated in the larger states and partly because of peculiarities in how Electoral College votes are apportioned. A person voting in Wyoming has 3.6 times as much influence on the Electoral College tally as someone voting in California. With time, demographic trends will make this imbalance even more unfair and destructive—as majority Democrats, increasingly concentrated in cities and larger states, are overruled by minority Republicans from rural areas and smaller states.[35]

The House of Representatives was meant to be the seat of the democratic principle—"one man, one vote." But, from the very beginning, political trickery made a mockery of noble sentiment. Blacks were conveniently not included, nor were women, or the unpropertied. And, from the start, the system was easily and enthusiastically gamed. "Gerrymander" is a hundred-year-old slang term derived by combining the last name of then Massachusetts governor Elbridge Gerry with the last half of the word *salamander* (the bizarre shape of a voting district he fashioned to gain electoral advantage for his party). Gerrymandering has since become a Republican Party art form, much aided by sophisticated computer technology that can sort voters block by block and aggregate them into the weirdest-looking voting districts—beyond the wildest dreams of the good governor. The GOP has also done its level best, by mostly foul means, to suppress the vote of those likely to vote against it. Dirty tricks include poor access to polling places, impediments to registration, purging voter rolls, voter intimidation, and photo ID requirements. These cynical measures disingenuously disguised as preventing voter fraud (which in fact has almost never been documented) instead fraudulently disenfranchise those who have fewer resources to vote.

Big money always has and always will stack the political deck—but the big money is bigger now than ever before and the stacking is greedier and more dangerous. The rich can get so much richer, at the expense of everyone else, partly because they use their wealth to buy politicians, who then slavishly pass laws allowing them to get even richer. This advantage to the advantaged occurred also in ancient Greece and Rome, in Mesopotamia and Persia, in India and China—everywhere in the world, in all times and in all places. But the combination of technology and tax policy has now given the United States the greatest wealth inequality in the world and one of the greatest in the history of the world. And big-money corporate interests were given an additional huge political gift when a GOP-friendly Supreme Court allowed unlimited campaign contributions, using the shallow excuse that it was protecting corporate free speech. Money buys lobbying power, money buys politicians, money buys votes, money buys the government.

The GOP also excels in guile and marketing. The Democratic Party has mostly good political values dragged down by lousy political skills. Republicans have mostly retrograde and unfair political values sold with slick public relations cynicism. They have systematically passed bills that benefit the big guys—providing tax cuts for the rich and corporate giveaways, while at the same time posing as the defender of the little guy. Instead of government of, by, and for the people, we have government of, by, and for the superrich individuals and corporations.

Societal delusions will be doubly hard to overcome because our election system is so systematically unfair and promotes, protects, and perpetuates the narrow interests of the very few over the welfare of the many. The will of the people, and the pressing needs of our future, are frustrated by big-money influence and bare-knuckle political manipulations. Because our system, in so many different ways, unfairly apportions political power to the minority, there is a systematic, one-sided, self-perpetuating barrier to the election of truthful candidates willing to confront societal delusions.

The first step up the steep hill facing the Democrats is establishing party unity, but here too the clear and systematic advantage goes to the Republicans. Psychological studies show that right-of-center voters are much more inclined to communalism and following the leader. Left-of-center voters are more often individualistic and as difficult to herd as cats. History has many examples of minority right-wing parties seizing autocratic power in a vacuum created by warring majority left-wing parties: Mussolini in Italy, Hitler in Germany, Franco in Spain, and Pinochet in Chile.

Trump seemed to be the most divisive and unpopular politician in Republican history, emerging badly bloodied from a primary battle with sixteen rivals that was long on brutal personal insults, on both sides. At no time did he command the loyalty of a majority of Republican voters, and for most of the primary he was bitterly opposed by about two-thirds. But once the general election began, Republican voters stepped into line and supported him with near unanimity, holding their noses and ignoring his lurid sexual and financial past, his bad manners, his subservience to Russia, and his often anti-Republican policies. They are still supporting him almost unanimously despite all the scandals and missteps of his early days in office.

The Democratic primary between Bernie Sanders and Hillary Clinton was a contrasting lesson in political civility—small policy differences discussed with elevated and polite discourse. But many Bernie backers were strident and persistent in opposing Clinton, and took their grudge past the primary and into Election Day. They encouraged Clinton voters to stay at home or vote for a fairly loony Green Party candidate, who took away a million Clinton votes that would have swung the election. I pleaded with left-of-center voters (in person, in blogs, and in tweets) not to repeat the terrible mistake of 2000, when Ralph Nader's well-meaning candidacy handed the election to a disastrous Bush presidency. But repeat it we did. Freud called it the "narcissism of small differences"—the tendency of the bitterest battles to be fought over the smallest of issues by those who are most alike. Democrats divide; Republicans coalesce. The Republican lock on power is a disaster for our country and the world, reinforcing every societal delusion and preventing any serious efforts to solve them. Democrats, also often under the thrall of selfish special interest groups, are very far from perfect. But, warts and all, they represent our only near-term hope for facing and solving our problems, and the world's. Making America great again means saving us from the disastrous GOP policies. And this requires a unified Democratic Party and one that is much better at delivering a meaningful and coherent message.

Campaign Skills (and Lack Thereof)

Trump's enormous success as a presidential campaigner and his disastrous failures as a president prove that there is little correlation in the skills required for each activity. This is especially dangerous in a make-believe, unreality-TV world that judges presidential candidates on stagecraft rather than statecraft. Clinton was an uninspiring candidate, but she would have made a good enough president. Trump was (to some) a thrilling candidate, but he is now perhaps the worst (and certainly the most unlikely) of all possible presidents.

In a postmortem, you examine the body after its death to determine what caused it. For many, Clinton's unexpected defeat felt like a sudden death in the family—not so much because she lost, but because Trump won. Doing a quick post-election review helps clarify what she did to lose (despite being a decent person and having so many seeming advantages), and what he did to win (despite being a terrible person and starting as such an underdog). Understanding how her campaign was so unexpectedly bad, and his so stunningly successful, may help explain what happened and prevent it from happening again.

The mystery is how Clinton, a very smart person, turned out to be such a political dunce, while ignorant Trump turned out to be a political genius. It starts with demeanor. Clinton had long fought for the little guy, and Trump had always fleeced him—but, on the stage, she was remote and he was immediate. She told the truth unconvincingly. He could lie, effusively, extravagantly, and effectively. Trump's voters bought the topsy-turvy counterfactual belief that he cared, she didn't; that he was trustworthy and she was not. And, mistakenly thinking she was far ahead and shouldn't risk looking bitchy, she let him get away with it. Of course, Clinton was fighting an uphill battle. For twenty-five years, she had been systematically, strategically, and viciously demonized by the radical right and the campaign against her became a full-court press after 2008, when her opponents anticipated correctly that her turn would come again in 2016. The incessant propaganda against her seduced many people into an instinctively amygdalar hatred, unsupported by any cortical rationality.

But Clinton did have a tin ear for the common man. Twenty years ago, when she first ran for the Senate, she went on a "listening tour" to almost every tiny town in New York State, making the point that she wasn't some city slicker who felt entitled to automatic victory. In this election, Clinton was remote and inaccessible, assuming she could rest comfortably on her long lead and past laurels. Rural and Rust Belt voters never felt she respected them or understood their plight. Real policy solutions come out of close contact with the people those policies are meant to serve. Without learning from, and about, the constituency you propose to serve, how can you hope to win their confidence?

Clinton's highly honed, Yale-trained rhetorical skills turned out to be a big handicap. She is a brilliant lawyer who speaks in an admirably coherent, well-modulated flow of perfectly constructed paragraphs. Trump is often incoherent, always inarticulate, and shines best when limited to 140-character tweets. But many in the crowd loved Trump and tuned out Clinton—because he had the common touch and spoke their language—and she didn't.

Trump could make himself sound as if he were addressing each person in the audience as an individual; connecting with them on a personal level; understanding and validating their pain, their fear, and their anger; that he was a Big Daddy ready, willing, and powerful enough to make things right. Skilled and experienced as a lifelong con man, Trump could, cynically and unashamedly, play on his supporters' trust with the false promise that he alone understood what needed to be done and he alone could make their lives better. It was all phony, but his supporters heard what they needed and wanted to hear, and could not distinguish his lies and fantasies from the world's realities. Clinton's speeches were right on message, wrong on tone. -

Clinton always seemed stiff and scripted; Trump was all un-bridled emotion. Her rallies were dull and poorly attended; his became love-ins. The emotions he expressed (and that some of his adoring admirers experienced) were often vile, but always cathartic—both for him and for them. Clinton failed to click emotionally, even with those groups who naturally should have been her strongest supporters—women, Latinos, blacks, Muslims, immigrants, and gays. How incredible that Trump won the votes of 53 percent of white women, despite being a shameless misogynist who brazenly bragged about serially forcing himself sexually on any attractive woman who took his fancy. Clinton, in contrast, was pathetically unable to mobilize the aspirations of her gender, even though, after centuries of women being treated as second-class citizens, one was finally poised to achieve the country's highest office. Clinton's voter turnout among Latinos and blacks was much less than Obama's despite Trump's constant threats against, and disparagement of, both groups and his close ties to white supremacists. Clinton always got the words just right, but she was tone deaf when it came to the music.

Clinton was all facts, no feeling; Trump was no facts or fake facts and all feeling. During the campaign, Clinton posted 65 fact sheets, totaling 112,735 words, describing her policies in exquisite detail on every conceivable topic. Trump barely bothered to have policies, posting just seven statements, totaling about 9,000 words—signifying nothing. His plea for support was framed exclusively in attack tweets and provocative slogans: "Make America Great Again," "Build the Wall," "Put Her in Jail," "Drain the Swamp." Dramatic (and inaccurate) images and metaphors can easily short-circuit (accurate) intellectual explanations of important issues and problems. Clinton won the policy debate, but lost the election. Her campaign was aimed at the mind. Trump aimed straight for the gut.

Clinton hated campaigning more than going to the dentist, and it showed. Trump loved and needed the admiration so much that his presidency has become a never-ending campaign tour. Clinton's considerable (and fully understandable) contempt for Trump also unfortunately extended to many of his followers, in a way that she could never completely control or hide. One turning point in the election was her being caught calling them (in a rare unguarded moment) "a basket of deplorables." Some of Trump's racist followers were indeed deplorable. But Clinton's inability to channel her feelings more selectively cut off any chance of capturing the support of the many less committed, more reluctant Trump voters. She lacked the political gift that her husband had displayed in such extravagant abundance but could not transfer to her. You felt that Bill Clinton connected with your pain. Hillary Clinton loved humanity, but didn't seem all that comfortable with individual people. Bill Clinton, in his prime, would have beaten Donald Trump in perhaps the biggest landslide in American history.

Trump gave the deceptive appearance of genuineness—despite telling an endless string of lies. Clinton was mostly truthful and sincere, but gave the false impression of being a phony. People believed Trump because he always seemed to say the very first thing that came to his mind, even though it was usually utterly false. Clinton seemed less sincere, even when being completely honest, because her manner was so studied. Trump seemed truthful when lying because it is his ingrained second nature.

So, how does it add up? Successful politicians succeed in winning hearts and minds because they are good at understanding and exploiting human nature. Platforms, policy, and pronouncements are just empty words without empathic connection—the bond that conveys to the electorate that they are listened to, understood, and will be cared for. Hucksters like Trump are expert at feigning this connection, to the detriment of all of us. True statesmen aspire to see political life as a selfless journey with, and for, their constituents—not a game of self-promotion and self-aggrandizement. At the beginning of our experiment in democracy, political discourse was conducted in the high-toned intellectual style of the enlightenment. Arguments used logic and were meant to appeal to reason. In this past election campaign, the contest between the cortex and the amygdala was won by the amygdala—extravagant emotion triumphed over rational thought.

Trump won. American democracy, societal sanity, and the future welfare of our children and planet all lost. He isn't crazy, but we are for electing him. And for allowing our society to degenerate to the point that someone like Trump could be taken seriously as a presidential candidate. The contingencies contributing to his win were crazily long-shot, requiring the unlikely convergence of sixteen Republican dwarfs in the primary, a vulnerable opponent in the election, the Putin push, the heavy hand of the FBI, the spite of Julian Assange, and wacky third-party candidates splitting the vote. We are now paying the price.

Chapter 5

Trump, Tribalism, and the Attack on Democracy

A dolf Hitler comparisons are properly denounced as clichéd and out of bounds. First ridiculed as *Reductio ad Hitlerum* by Leo Strauss in the 1950s, the phenomenon has been rechristened Godwin's law by the Internet age: "As an online discussion grows longer, regardless of its topic or scope, the probability of a reference or comparison to Hitler or Nazis approaches 1."[1] But recently the meme's creator Mike Godwin has moderated his objections. Flooded with requests for advice on acceptable Internet behavior given that some people and situations now bear more than a passing resemblance to 1930s Germany, his reply is liberating: "Let me get this Donald Trump issue out of the way: If you're thoughtful about it and show some real awareness of history, go ahead and refer to Hitler or Nazis when you talk about Trump." Hitler/Trump parallels would be cheap shots if Trump hadn't brought the lines so close. We must learn from Hitler's takeover if we are to prevent Trump from making it happen here. And we must push back against him immediately and in force, before it is too late to push back at all.

Hitler, like Trump, never won a popular election—his best performance at the polls garnered only 44 percent of the vote. Hitler, like Trump, had only the greatest contempt for democratic traditions, a free press, the courts, intellectuals, human rights. Hitler, like Trump, regarded truth as negotiable, lies as effective weapons, and morality as excess baggage. Hitler, like Trump, was a conspiracy theorist who surrounded himself with subservient "yes men," unwilling or unable to challenge his misconceptions and misjudgments. Hitler, like Trump, was a world-class narcissist. Hitler, like Trump, was despised and underestimated by the political establishment, who felt he could be used and manipulated for their own purposes. Hitler, like Trump, defied the political establishment and remained true (only) to himself. Hitler, like Trump, felt disrespected and treated unfairly, and had many scores to settle. Hitler, like Trump, claimed infallibility, that he was smarter than his generals and advisors, and that his gut instincts were the nation's best guide. Hitler, like Trump, exploited the fear, anger, and resentments of his people. Hitler, like Trump, promoted tribalism and reviled minorities as dangerous vermin. For sure, Hitler was also unlike Trump in some ways. He was much smarter, better read, more mature, better organized, less ignorant of history, more self-disciplined, less distractible, better mannered, less needy, more plausible—and, so far, much more bloodthirsty, ruthless, and deadly.

Hitler's seizure of dictatorial power was simple, swift, and complete—exploiting a mere constitutional formality to destroy all constitutional restraints on his authority. It was so very easy, so seemingly innocent, so sudden, so irrevocable, so devastating. The vehicle was the euphemistically named Enabling Act—an amendment to the 1933 Weimar Constitution that gave Hitler "plenary power" to enact laws without needing the approval of his parliament or judiciary. This allowed him to abolish civil rights and kill off his opposition with no muss, no fuss—it was perfectly legal and required only a cowed parliament and a few strokes of the pen. The scariest lesson for us is that one terrorist act, the Reichstag fire, was sufficient trigger for Hitler's all-powerful dictatorship. "National security" is an effective and convenient excuse for destroying democracy. We will surely suffer more terrorist acts. Trump will seize on them—if we let him.

The analogies between Hitler and Trump are all too obvious. Thankfully there are also some obvious differences. We are not as fragile as Weimar Germany—our constitutional democracy has the solid tradition of almost 250 years behind it, not the mere 14 years of Weimar Germany. We are not in a devastating economic depression with unemployment at 30 percent and a startling inflation rate of 1 trillion percent over just one generation. We haven't been defeated and humiliated in the world's most destructive war. We haven't suffered millions of war deaths and crippling casualties. We don't have a grinding, daily and deadly, civil war on the streets between the radical elements of the left and right. But it can't be reassuring that, even given our privileged position, Trump has come so far, so fast, and has such a clear course to subverting our democracy. He can be stopped if there is a sustained opposition from Congress, the courts, the media, and the people. But so far Congress smells of Weimar, the courts are untested and will swing further right with Trump's appointments, the media is taking heavy fire, and the people are only beginning to awaken.

The Good, the Bad, and the Ugly

In the midst of living it, you can never know exactly how history will play itself out. The contingencies are so numerous, interacting, and unmeasurable that seemingly small events can have surprisingly large impacts. Doubly true at this unpredictable tipping point—the first time our democracy has ever faced a possible takeover by a determined demagogue. When I'm doing psychotherapy, it's useful to ask the patient to dope out his hoped-for best outcome, his most feared outcome, and the possible outcomes in between. People who consider their possible futures have better futures—less improvised, more within their control. Intelligence gathering reduces surprises and avoids blindsiding and wrongfooting. This is perhaps why dystopia books suddenly became bestsellers after Trump's inauguration. People are trying to understand their scary present to avoid an even more frightening future—feel-good, escapist books are less relevant in an environment of threat. This may be a promising sign that we are willing to face reality and drop societal delusions.

Let's play our own futuristic guessing game—imagining the worst, the best, and the middling future outcomes of Trump's presidency.

Worst-Case Scenario: Il Duce Trump

It started when Trump tweeted that the press is not just his enemy, it's the enemy of the American people—and began calling reporters the "opposition party." His press secretary divides and conquers by giving friendly press juicy exclusives and barring the truthful press from attending briefings—something never done before. Trump keeps spewing a steady stream of fantastic lies; the press keeps up a steady stream of persistent fact checking. His base, 40 percent of the country, continues to believe Trump, or at least is willing to give him the benefit of the doubt. The truth gets so obscured in the fog of obfuscation that many stop thinking it is all that important. Trump also barrages the courts, challenging judicial review of executive orders related to "national security." He gradually wears down resistance to the "Trump doctrine" that protecting against terrorism is more important than protecting civil rights.

The federal government is rapidly "deconstructed," partly by design and partly because Trump's appointees are so incompetent and ideological that many of the most talented government employees resign—or are fired—rather than work for him. White House policies are constantly changing and their implementation is inconsistent. Hundreds of major government leadership positions remain unfilled because most potential recruits are unwilling to work in a Trump administration. The Environmental Protection Agency is gutted, the Paris agreement cancelled, and the energy industry deregulated. Corporations, reveling in the lack of government oversight, reduce attention to product and worker safety, lower wages, and raise prices. Corporate executives dislike and fear Trump, but support whatever he does because it is good for business and may earn them giant tax cuts.

The FBI's investigation of Russian election tampering is obstructed when Trump fires the FBI director and is delayed further as he refuses to cooperate or accept the findings of the independent Special Counsel. Trump's mounting frustration with investigations and criticism results in his posting constant barrages of inflammatory and internally inconsistent tweets attacking the press, his political opponents, formerly friendly foreign leaders, leaders of his own party, and even members of his inner circle. Frequent White House reshuffles, with many people fired or reassigned, result from recurring staff power struggles. Trump's statements and decisions flip-flop wildly depending on which advisor currently has his ear or which rumor is trending on right-wing social media. It turns out that many of these have been planted by Putin's disinformation team.

Under daily attack, Trump uses the national security excuse to extend his powers and erode constitutional restraints. He issues an executive order ending all the Russian investigations and sealing records, citing their "potential to compromise our nation's unity and give aid and comfort to the enemy." Large protest demonstrations in fifty American cities result in numerous arrests. To help restore order, temporary censorship is enforced on the mainstream media that is blamed for fomenting the disorder. As a further distraction, Trump instructs local police to round up tens of thousands of undocumented aliens.

When the inevitable next terrorist event occurs on U.S. soil, Trump declares a state of "national emergency" and blames the courts for previously not giving him sufficient powers to prevent terrorism. He vows never again to be deterred from protecting America because of outdated and now dangerous judicial checks and balances. The protest demonstrations escalate and are countered by pro-Trump private armed militias. A spree of retaliatory hate crimes throws many cities into turmoil. Martial law is declared in communities with heavy Muslim populations and thousands of young Muslim men are "preventively detained" on suspicion of jihadi sympathies. Some are tortured. Repeated protest marches continue to roil American cities—some opposing specific Trump actions; others, his assumption of autocratic powers. Some are peaceful, some are rowdy. Some lead to violent clashes with angry and well-organized Trump supporters. Many are broken up by police using tear gas, dogs, clubs—following Trump orders to get tough at any sign of public disorder. The press is enjoined from covering demonstrations, on grounds of national security. Thousands are arrested, hundreds are badly injured. A second wave of protests is met even more brutally. Public assembly is temporarily prohibited, but the emergency prohibition is never lifted.

Trump warns the American people that foreign and domestic enemies are hoping to bring the country to its knees with anarchic acts of violence. He promises to protect America at any cost and to "make it great again." Congress and the Supreme Court are cowed. Opposition to Trump's takeover is muted and ineffective. The Democrats lose heart and are divided on what to do next. The press is thoroughly censored and dozens of journalists are arrested and charged with treason. Social networks are muzzled and directed toward supporting a strong central response to the perceived threat.[2] The military and police follow orders. Trump's temporary powers are extended indefinitely because there is no natural end to the emergency. Trump is the first American dictator, but not the last. After Trump dies suddenly (conspiracy theorists suspect poison), a succession power struggle among his children and aides is won by son-in-law Jared Kushner, who rules with Ivanka at his side.

Best-Case Scenario: Democracy Defends Itself

There is no Trump post-election honeymoon. The press refuses to be muzzled, and people quickly wise up to his increasingly implausible lies, erratic behavior, and unmet false promises. Trump's approval ratings, already abysmal at Inauguration, drop to the mid-twenties to low thirties. Internet-based populist groups organize frequent and large anti-Trump rallies to put persistent and powerful pressure on politicians to oppose his legislation. Key Republicans who had cynically hoped to ride Trump to victory in a Tea Party coup now realize cynically that their survival depends on getting off his sinking ship. It turns out that just twenty-seven people can stop Trump. Three senators and twenty-three members of the House of Representatives (a few motivated by patriotism, others by political calculation) break ranks with the GOP leadership, giving the anti-Trump, pro-democracy coalition a stable congressional majority that repeatedly opposes all of Trump's attempts at power grabbing. One GOP-appointed Supreme Court justice joins ranks with his Democrat-appointed colleagues to refute the constitutionality of Trump's dictatorial executive orders. A gaggle of governors assert the right of states to follow the Constitution, rather than Trump's gut impulses. Trump's base dwindles down further to extremist die-hards, as everyone else realizes that his policies are disastrous and his person is despicable—a would-be emperor without clothes. He backs down, stops issuing executive orders, and restricts himself instead to bitter tweeting and constant campaign events with an ever-shrinking and less adoring base. Democrats take over the House of Representatives in 2018 and win Senate seats in states that were previously GOP strongholds. In 2020, Trump is replaced in a landslide by a unifying president who uses the public's mandate and the Democrat-controlled Congress to pass a legislative program that heals the wounds, abandons societal delusions, and meets the challenges of the future.

Most Likely Scenario: We Muddle Through

Trump's presidency is the most divisive in American history, except for Lincoln's. His governance, both of himself and the country, is chaotic and unpredictable. After eight years of no-drama Obama, the country is rocked by daily crises, frequent resignations and purges of people in high places, peaceful anti-Trump demonstrations and rowdy pro-Trump counterdemonstrations, government agency breakdowns, and resignations of national security and military leaders. Government agencies lose credibility and are run into the ground. The Republican Congress mostly sticks with Trump and passes lots of retrograde legislation that further disadvantages the disadvantaged and puts the world on an even greasier slope toward global warming. Health care, a disorganized, inequitable, and expensive mess before Obamacare, becomes even worse after major parts of it are repealed. The Supreme Court mostly goes against Trump on major constitutional issues, but allows him to gradually erode civil rights, voter enfranchisement, and corporate regulation.

As Trump becomes more and more frustrated by falling poll numbers, constant attacks, and GOP defections, he gets increasingly bitter and insulting, even to those who have been mostly loyal supporters. Tweets become more numerous, more conspiratorial, and more laughable. The GOP gets punished in 2018 and especially in 2020, but a lot of the damage is already done and hard to fix. Two thousand twenty is the biggest Democratic landslide in history as voters express their buyers' remorse for ever allowing Trump into the White House. Historians form a consensus that he is America's worst president and ponder the underlying social, political, economic, technological, and demographic forces throwing up such a terrible leader. Many academic papers are written on the joys and perils of democracy. Dystopic books and movies continue to be popular.

Color War Threatens Democracy

I first started going to summer camp as a preteen, in the 1950s. I loved the nature, the camaraderie, the sports, and learning new stuff. But I hated Color War. The camp was arbitrarily divided into the Blues, the Reds, the Greens, and the Yellows. We were pitted against one another all summer in every conceivable (and lots of inconceivably silly) challenge tests. Many of the guys took seriously the color loyalty, the competitions, and the awards. But it just seemed dumb to me—especially since I didn't particularly like many of my fellow Greens, and the kids I did like had been assigned to the other colors. It raised the important question of whether my then hero worship of Mickey Mantle and the New York Yankees made all that much sense—perhaps the hated Dodgers were nicer people. Years later, I had the same cycle of disillusionment with my college fraternity—it was a great relief to have a supportive tribe as a freshman, fun as a sophomore, constricting as a junior, and completely irrelevant as a senior. As Einstein recognized, tribalism is an immaturity you can outgrow.

I recently came across the classic Robbers Cave study of "color war"—published in 1954, coincident with my camp summer and also with the publication of *Lord of the Flies* (the ultimate "color war" novel). The experiment stands up as a powerful metaphor for the tribalism tearing our world apart—and as a practical guide on ways to end it. Two groups of fifth-grade boys were invited to a "summer camp" experience in the mountains of southeast Oklahoma. All were middle class, Protestant, from the same geographical area, psychologically intact, and with above-average intellectual functioning. Each group first participated in one week of camp activities isolated from the other group. They each spontaneously formed a cohesive identity, even giving themselves totem names (the Eagles and the Rattlers). When the groups were then allowed contact with each other, an "us" versus "them," competitive attitude quickly formed. The counselors responded by setting up games that involved valued rewards and trophies. The two groups started skirmishing over issues large and small and competition became especially fierce when resources were made scarce (as when camp dinner food ran out before one of the groups was called to eat). Sports competition released a trash-talking, stereotyping exchange of insults. Before long, the groups were raiding each other's bunks, destroying each other's property, stealing prizes, burning each other's flags, making threats, and planning to commit acts of outright aggression. Color War had become literal war—stopped only when the counselors intervened. *Lord of the Flies* in the Oklahoma hills.

Attempting to reverse the animus, the counselors brought the groups together in a variety of noncompetitive pursuits: eating together in the dining room, going on outings together, doing chores together. Mutual dislike and unwillingness to comingle persisted. Cohesion across groups emerged only when they were forced to confront a series of study-rigged "mishaps" requiring them to work together and endure mutual sacrifice. The warring factions merged when mutual interest became more important than prior tribal difference. This led to an unexpectedly happy ending: when one of the two groups won a cash reward at the end of camp, it opted to share with the other, so that everyone could partake in a parting round of malteds.[3]

The science in this study and the art in *Lord of the Flies* both describe the same spontaneous emergence of primitive tribal aggression—something that seems engraved in our social DNA. The bad news is that we will so easily do simply terrible things in the seeming good cause of tribal loyalty. The good news is that intergroup animosity can be diminished when people must rely on one another to meet mutual challenges or face a common enemy. Unfortunately, it is much easier to create rivalry than to resolve it—but, fortunately, under the right circumstances, cooperation can replace competition.

Color war tribalism is a ubiquitous, dispiriting part of modern life—increasingly violent because population pressure is increasing and resources are becoming scarcer. It's why Shiites are killing Sunnis and Sunnis are killing Shiites; why Israelis and Palestinians have engaged in a seventy-year peace process that never brings peace; why the "Reds" fought the "Whites" in the Russian Civil War; why the "Blues" fought the "Grays" in ours; why we wave flags, root for sports teams, love our country; why whites have been so intolerant of all the other shades of human color; why some groups circumcise and others don't. And also why Republicans (dominating the "Red" states) and Democrats (dominating the "Blue") are currently having so much trouble finding common ground to solve national problems.

A few years ago, my best friend and I were watching a simulated "color war," artificially created as shtick in a SeaWorld maritime show. The whole thing took thirty ridiculous minutes. Two teams—Red and Blue—competed in ten events that included boating, waterskiing, swimming, biking, and a series of highly contrived and fiendishly foolish stunts. The master of ceremonies arbitrarily divided the audience of two thousand into alternating sections of about one hundred, instructing each to root en bloc either for the Red or for the Blue. Within a few minutes the sections were enthusiastically cheering on their assigned teams. By the end of the show, some people were screaming loud insults at members of the opposing team and at its supporters in nearby sections. My friend turned to me sadly and said, "Is this the way it had to be?" So far, yes.

Human tribalism once had great evolutionary survival value. Our hunting and gathering ancestors were completely dependent on their small group, economically and for security, and would die almost immediately if banished or separated from it. But, living now on a shrunken earth, tribalism may be the most lethal baggage we carry forward from our past into our uncertain future. And the Trump effect is increasing it throughout our society, most disturbingly in our children. A friend of mine who lives in the American South told me the sad story that his five-year-old son and another little kid got into a fistfight in their kindergarten class, one defending, the other attacking Trump. My observant granddaughter, a high school sophomore, has also noted a dramatic change since Trump was elected. Previously, class discussions were civil and the views expressed spanned the political spectrum. Now, emboldened by Trump's victory, class and schoolyard discussions are acrimonious and dominated by kids willing to assert racist, sexist, xenophobic, and antigay remarks, previously considered taboo. Her interpretation is that the kids haven't changed overnight, but the standards of civil discourse have. Pollsters underestimated Trump's appeal to people before the election precisely because some of those surveyed were too ashamed to admit they would vote for him. Now they, and their children, have come out of the closet.[4]

To reverse the polarization of our society (and, with it, the erosion of our democracy) we must bridge the ever-widening tribal gap between "us" and "them." Not since the Civil War have the two poles of the political spectrum become so alien to each other. Many left-leaning urbanites have written off Trump's base as know-nothing, head-in-the-sand bigots and boobs. They, in turn, view liberals as elitist, condescending, soft on terrorism, naïve, "tax-and-spend liberals." None of these epithets, on either side, are all that accurate and they certainly don't advance discourse toward common understanding. Part of our polarization reflects substance; much of it reflects political manipulation and rhetorical sleight of hand. We fail to work out mutually satisfactory solutions when we talk past each other, insult each other, use confusing language, or stop listening. We must set aside our partisan jargon and speak in plain language about what works and what doesn't.

Psychological Weapons Win Political Battles

J. Robert Oppenheimer, father of the atomic bomb, had creator's remorse: "The physicists have known sin: and this is a knowledge they must not lose." Psychology has also known sin—we helped create the noxious weapons used in the political propaganda that has so brutalized and cheapened our democracy.

Advertising is the art of tricking people into buying things they otherwise might not want or need. Political advertising is the art of selling people bad ideas and getting them to support politicians who don't have their best interests at heart. Advertising is applied psychology—it works by bypassing our conscious cortical reasoning processes, in order to manipulate our unconscious amygdalar emotions. The explosion of psychological theories in the late nineteenth century (psychoanalysis, behaviorism, and social psychology) almost immediately led to their use (some might say misuse) in the peddling of consumer goods. In the last few decades, psychology has been used (and, in this case, definitely misused) in peddling political snake oil.

Edward Bernays is known as the father of public relations, a term he invented because it sounded so much more genteel than what it had previously been called (and really was): propaganda. He happened to be Sigmund Freud's nephew and made his fortune combining techniques derived from psychoanalysis, behaviorism, and group psychology to improve corporate balance sheets. His basic insight: "If we understand the mechanism and motives of the group mind, is it not possible to control and regiment the masses according to our will without their knowing about it?" This led to his special expertise: the influencing of consumer behavior by what he called the "engineering of consent." Bernays pioneered the mass marketing of fashion, food, soap, cigarettes, books, and a multitude of other consumer products. Under his clever stage-managing, the sight of a woman smoking in public lost its whiff of immorality and instead became fashionable, politically correct, and appropriately sexy. All it took was suggesting that cigarette packages be color-coordinated with each year's fashion colors and arranging the 1929 New York City Easter parade as a showcase for beautiful models holding Lucky Strikes (labeled as little "Torches of Freedom").

Bernays also invented the concepts of celebrity and thought-leader endorsement: "If you can influence the leaders, either with or without their conscious cooperation, you automatically influence the group which they sway." Bernays's marketing brilliance is still felt in our everyday breakfast behavior. Working for the pork industry, he surveyed five thousand physicians and publicized widely the result that a heavy bacon-and-egg breakfast is far healthier than the then customary light breakfast of tea and toast. Of course, this was complete nonsense, but the message stuck. Bernays and P. T. Barnum were kindred spirits, each believing there is an easily snookered sucker born every minute. Both got rich banking on it. As have the multinational corporations. So too the big political players.

At about the same time, John Watson also struck it rich translating psychological theories into advertising gold. His was a rags-to-riches story possible only in America. Poor but ambitious boy gets good education, works his way up to become America's most eminent psychologist, then suddenly throws it all away to start a new career making a fortune as a top executive in the newly burgeoning advertising industry. Extending Pavlov's conditioning of dogs to humans, he realized that human behavior could be greatly influenced by subliminal methods that bypass conscious awareness. He called the approach "behaviorism" because it had no interest in (or appreciation for) the complexity of consciousness and the human mind; humans and dogs were equally manipulable. Watson used his methods of behavioral control to persuade people to buy products. He invented the coffee break to push Maxwell House coffee, persuaded women that smoking was sexy so long as they used Pebeco toothpaste, and convinced them of the need for a cabinet full of beauty products. It's a frightening juxtaposition, Watson as father of both behavioral psychology and modern advertising. He brought science to consumerism.

The methods developed for consumer advertising have played all too well in the even dirtier world of political propaganda. Joseph Goebbels, Hitler's mouthpiece, never got a degree in psychology, but he did study it carefully. Bernays lamented: "Goebbels was using my book *Crystallizing Public Opinion* as a basis for his destructive campaign against the Jews of Germany. This shocked me." Foreshadowing Trump's campaign strategy, Goebbels once said: "There was no point in seeking to convert the intellectuals. . . . For the man in the street, arguments must therefore be crude, clear and forcible, and appeal to emotions and instincts, not the intellect. Truth was unimportant and entirely subordinate to tactics and psychology." This about sums up the Republican Party's protracted and deceitful campaign to discredit Hillary Clinton.

Political advertising in the United States came of age with TV. In the presidential campaign of 1948, Harry Truman shook a million hands and traveled 31,000 miles. It was a lot easier for Dwight Eisenhower in 1952. He filmed 40 TV spots in one day that covered much greater territory. John Kennedy, our most telegenic president, spread his message in 200 TV commercials. The full potential of negative advertising became obvious in 1964, when Lyndon Johnson's "Daisy" ad depicted Barry Goldwater as an unreliable trigger finger, eager to start the nuclear apocalypse. Most political propaganda since then has been negative, not offering positions on what should be done but rather attacking the opponent's policies and often his person. Talk radio degraded political discourse into coarse conspiracy theory. The Internet helped lies travel across the world in seconds.

Stereotyping is key to political polarization and brainwashing. *Webster's* defines a "stereotype" as a "standardized mental picture that is held in common by members of a group and that represents an oversimplified opinion, prejudiced attitude, or uncritical judgment." The word *stereotype* is only two hundred years old, comes from the French, and originally described a printer's mold used to produce standardized copies. But the tendency to stereotype is as old as our brains. It provides a quick shorthand to understand experience and establish group membership. Once you develop a preconceived notion, it's hard to change, especially if it's shared by fellow group members. The vicious cycle: polarization between groups causes stereotyping, which increases polarization, which causes more stereotyping. Inadequate knowledge and understanding of people different from us is fertile ground for negative stereotyping, especially when politicians cynically pit us against one another—always for their advantage, never for ours. Stereotyping is the most efficient way of dealing with simple problems, but the most destructive way of dealing with complex ones.

Politics, like most professions, creates its own language, using shorthand buzzwords to impede rational thinking and civil discussion. A "buzzword" is defined as an "important-sounding word or phrase, often of little meaning, used chiefly to impress." Gresham's law in economics says that "bad money drives out the good." Applied to politics, Gresham might have it that meaningless buzzwords drive out rational thought and meaningful discussion. They short-circuit dialogue, appealing to the amygdala while keeping the cortex at bay—a lazy substitute for thinking, deliberately disguising and obfuscating to derail common sense and good judgment. Imprecise and purposely vague creatures of fashion, jargony innuendos are usually invented to make a propaganda point rather than to seek the truth or present viable options. The words used to frame political issues keep us from resolving them and hide self-interest and prejudice. The Republican Party's buzzword "tax revolt" (which sounds populist) is a disguised way of allowing Trump and his billionaire buddies to avoid paying their taxes.

Each color-war political party has its favorites, but the GOP is much cleverer at coining them and much more successful at imprinting them on the national psyche.

Republican buzzwords: free market/right to life/entitlements/class envy/states' rights/activist judges/death panels/welfare queens/liberal elite/latte liberals/limousine liberals/tax relief/war on terror/anchor babies/communist/socialist agenda/birther/safety in the streets/national security/patriotism/Freedom/Liberty/personal responsibility/Founding Fathers/religious freedom/the elites/the establishment/tax and spend/politically correct/originalist

Democratic buzzwords: fairness/fascist/green/islamophobia/leveling the playing field/paranoid/redneck/social justice/

sustainability/blaming the victim/civil liberties/choice/disad-
vantaged/diversity/equality/extremist/radical right/marriage
equality/misogynist/trickle down/narcissist[5]

"The Extremist Effect" describes the powerful propaganda
strategy of calling someone an "extremist" even if what he is
saying is perfectly mainstream and commonsensical. Someone
expressing middle-of-the-road support for women's rights can be
discredited as a "feminist" (or, if you are Rush Limbaugh, as a
"feminazi"). Preferring not to have the environment degraded by
global warming or pollution is discounted with the now strangely
pejorative term "environmentalist" (if you really want to make
them look foolish, call them an "extremist environmentalist").
And "liberal," which, not so long ago, was for many a term of
considerable pride, has now somehow become an insult, con-
noting moral laxity and economic naiveté. You don't have to
counter a position with contrary evidence, just vilify the person
making it.

If there are any Trump fans reading this book, they can rightly
accuse me of repeatedly and enthusiastically taking advantage of
"the Extremist Effect." My only excuse is that Trump is about as
extreme as any American politician has ever been, and sometimes
you need fire to fight fire. Truth is not always in the middle of
any two opposing arguments—and it is never in the middle when
one side espouses an extreme view. The strategy of the Republi-
can Party, since its takeover by the radical right, has been to start
with, and stick to, inflexible and one-sided positions. This has
worked brilliantly. Democrats are usually less ideological, more
flexible, more willing to compromise, and more open to all prac-
tical solutions. Most Republicans were more like this before the
party was shanghaied by the fake populism of plutocrats. We will
be able to cure societal delusions only when we recognize that
facts are facts, scientific evidence is scientific evidence, and "alter-
nate" realities are self-serving lies, and not realities at all.

Goldilocks Government

There are three models of government—all containing a kernel of truth, all dangerous when taken to extremes. Democrats tend to see government as a nurturing mother, doing good things for the people, taking care of their needs. Republicans want a Strong Daddy government—militarily powerful to stand down foreign enemies, strict on law-and-order measures to restrain and punish terrorists, criminals, and people who don't share their religious and moral beliefs. Libertarians see government as a nagging and intruding nanny—they believe a government governs best when it governs least. Political posturing polarizing these positions obscures the need for balanced Goldilocks solutions. Government must meet certain needs that can't be met in other ways. Government must establish a safe environment. Government must set the rules of the game and be an impartial umpire to ensure it is played fairly. But it must also leave the players free to enjoy playing and make the most of their abilities.

Trump is "deconstructing our government" by employing an army of incompetent and ideological leaders intent on destroying the agencies they have been charged to lead. The Environmental Protection Agency will no longer protect the environment. The Justice Department will no longer ensure we are just. The Department of Energy is run by the Energy Industry. The Department of Health and Human Services denies people medical coverage and makes its services less and less humane. The Food and Drug Administration is run by Big Pharma. The Department of the Interior intends to give away federal lands, not protect them. The Department of Housing and Urban Development aims to reduce subsidized housing. Our trade representatives are to abrogate the existing trade treaties. Our State Department is firing its diplomats. And to ensure that his appointees are enthusiastic in their hatchet work, Trump has appointed KGB-type attack dogs to each government department. Their job is to bully and spy on their "bosses" and report back at White House headquarters.

Trump has declared war against even the most necessary Good Mother government services and at the same time is attempting to seize unprecedented Strict Daddy powers, at the expense of our civil liberties. He is digging deep into government agencies to politicize their operations, attempting to destroy the civil service, which was introduced after the Civil War precisely to eliminate his kind of cronyism, conflict of interest, and corruption. Trump is taking power away from the servants of the people and concentrating it instead in his own hands and in the boardrooms of the multinational corporations. Not coincidentally, this was the strategy Hitler used—absolute political and police power to himself, almost unchecked economic power to Krupp and the other giant corporations; no power to the people.

Let's get this straight. I am a penny-pincher; I hate waste and want a lean and efficient government. But, that said, we have to face the fact that previous privatizations of what once were government functions have almost always been a colossal flop— there are some public services that get really loused up when done privately and for profit. Outsourcing is a classic mismatch of wonderful theory and disastrous practice. The argument for privatization seems compelling. Government is inherently bloated, lazy, wasteful, dumb, and inefficient because it does not have to face the discipline of the marketplace. Put public services up for private bidding and you will get the lower costs and greater efficiency that come with free market competition. But privatization practice usually results in a somewhat inefficient government monopoly being replaced by an even more inefficient (and often extremely greedy) private monopoly that is more expensive, wasteful, and lacking in accountability or responsibility for serving the public good. The selection of private contractors is usually rife with the corruption of political sweetheart deals. The profit motive consistently trumps public interest. And shareholders and executives benefit at public expense, while public services deteriorate. Let's do a quick review of the scorecard:

Infrastructure: Government does a lot better job than private industry in funding a whole range of activities that benefit the public: police, scientific research, roads, bridges, communication systems, utilities, sewers, flood control, national park and wilderness management. The area I am most familiar with is medical research. The findings that have most advanced health care come from the National Institutes of Health, not the pharmaceutical industry—whose "research" is devoted to earning obscene profits rather than helping patients. Tens of billions of dollars are wasted every year creating "me-too" drugs that serve only to extend patent life and allow for monopoly pricing. Would anyone want our Interstate Highway System to be privatized? Would anyone but Betsy DeVos, Trump's secretary of education, want to end our treasured system of free public education? Would we feel safe if all police forces were private and served only those who could afford to pay for them?

Defense: We pay outrageously padded bills to private military contractors who deliver poor services with constant cost overruns and virtually no accountability. Today's general becomes tomorrow's Raytheon vice president, negotiating sweetheart contracts with his former colleagues.

Medical Care: Our chaotic, confusing, profit-driven system is much more costly, wasteful, and less effective than the simple single-payer systems in most other developed countries, which can negotiate much lower prices and provide more accountable care.

Mental Health: Deinstitutionalization and the privatization of community mental health centers allowed the states to offload responsibility for the severely mentally ill so that now about 350,000 are in prison and 250,000 are homeless.

Water: Privatizing this precious commodity has made water more expensive and private equity funds rich.

Prisons: Privatizing helped make prisons one of America's big growth industries. Lobbying for unrealistically draconian drug laws has kept the lucrative contracts coming at enormous public expense and judicial inequity.

Courts: Chronic public underfunding has led to a kind of private entrepreneurship extorting large court fees that keep defendants in constant debt. You have to buy justice now in America.

Police: The frequent inappropriate police shootings are no accident—police are understaffed, underpaid, undertrained, and underscreened. A demoralized and dangerous police force is a disaster for poorer communities that depend on police protection, no problem at all for gated communities with a private security force.

Schools: Charter schools, once a great hope to shake up our moribund educational bureaucracy, have so far failed to live up to their promise and seem destined to benefit shareholders more than kids.

The consistent failures of privatization are not self-correcting. Privatization has a powerful political and economic momentum that defies logic, is immune to the consequences of repeated failure, is opaque to public scrutiny, and strenuously resists reform. What propels privatization, despite its obvious disadvantages? You guessed it—money doesn't just talk, it shouts. The profit motive can be very motivating.

Enormous campaign contributions from the superrich and big corporations (expressing their Supreme Court–protected, *Citizens United* right to free speech) promote the friendly politicians who support giveaway privatization. And there is a revolving door between government jobs and industry lobbying jobs that ensures sweetheart statutes and regulations that benefit the private contractors and harm the public interest. Behind it all are the greedy folks with the really big bucks who are selfishly eager to reduce tax-supported public services because they don't need or use them.

Capitalism and private corporations were invented in the Western world four hundred years ago as a response to flourishing trade opportunities. We have since accumulated a vast experience on its pluses and minuses and on the best-balanced relationship between public and private delivery of services. "Free market" Republicans look to Adam Smith as the father of their political philosophy, but he would be appalled at their misreading and misrepresentation of his intent. Smith pointed out the irreplaceable value of the free market in providing rational pricing and an efficient allocation of goods, services, and resources. But he also strongly supported the role of government in providing services that the free market could not: national defense, post office, police, firefighting, public works, health, education, justice, transportation, banking, controlling monopolies, enforcing contracts, and caring for the poor and infirm. As Adam Smith predicted 250 years ago, unbalanced systems don't work. Top-down, government-directed economies and free-for-all free-market economies both result in rampant corruption and misallocation of resources. The best-governed countries in the world, Germany and those of Scandinavia, have an appropriate balance of private and public services.

The excesses of unfettered crony capitalism in the United States were tamed, ironically enough, by a hero of the Republican Party, Theodore Roosevelt, who initiated the system of government regulation and national parks that Trump is now in the process of destroying. His "deconstruction of government" is a radical and impulsive gutting of institutions, built up over the past century to protect the public and our resources from corporate piracy. We must return to Adam Smith's balanced economy and to Theodore Roosevelt's fairly regulated one. Government should nurture, but not too much. It should keep us safe, but not by being a bully to other countries and to our own people. And it should give us as much liberty as possible so long as we are not infringing on other people's liberties. Government of, by, and for all of our people—not government of, by, and for a small circle of autocrats, billionaires, and multinationals. We are not equal in endowments, ambitions, opportunities, energy, or luck. Some people will always have more than others. But we should all play by the rules and have a fair umpire. Government can never outsource this responsibility.

No Crying Over Spilled Milk

Trump, the GOP, the corporations, and the billionaires won. America, ordinary people, democracy, and global climate lost. Trump, ever the raging bull in a surprisingly fragile china shop, has already done immeasurable damage and his carnage has only just begun. It will take the perspective of decades before we can evaluate the degree to which the harms caused by him are reparable or irreparable, but we can't waste a moment before initiating the most vigorous possible damage control. When my grandkids make a mistake in sports or school, my response is always, "next play." You learn from your mistakes, but you don't mope over them. Our next chapter outlines a next-play strategy for turning the seeming Trump disaster into a game-saving positive—a just-in-time chance to heal polarization and cure our society's insanity.

Chapter 6

Defending Democracy:
The Path Forward

In despairing moments, I entertain the thought that Trump must represent the probability gods' idea of a cruel joke. Or he may be the inevitable stigmata of mankind's evolutionary unfitness. Perhaps the game is already over and we just don't know it yet. Possibly so, but I think probably not. Sometimes it is darkest before the dawn and passivity, born of fatalism, leads to horribly self-fulfilling prophecies. We must shape our future, not hide from it on the loser's bet that there won't be a future to shape. Hope is both necessary and realistic. The depths of human stupidity and the heights of human resourcefulness tend to run in alternating cycles.

With Trump, our nation has hopefully shown its worst possible colors (barring a further regression to fascism or a catastrophic war or environmental disaster). He is a distillation, mouthpiece, and terrifying living embodiment of all the worst in human nature and societal delusion. If you were assigned the task of punishing humanity for its original sins, you could do no better than invent a Donald Trump and give him extraordinary power over the world's future. This will likely be one of the scariest four years in U.S. history; right up there with the War of 1812, the Civil War, the two world wars, and the Great Depression.

But disasters also offer opportunities and we are due for a rousing comeback. My best results as a psychiatrist were with patients in crisis. When someone hits bottom, there's no place to go but up—and bottoms are terrific motivators of change. Some of my friends voted for Trump in, what seemed to me, the strange hope that he would be an agent of "creative destruction." They may, surprisingly, turn out to be right, but not in the way they meant it. His destructiveness may be a perfectly timed shock treatment, necessary to put us in touch with the realities of our world and cure our societal delusions. As people come to see through Trump, it is equally important they see through the naked self-serving policies of the Tea Party he represents. Trump in office is better than Trump impeached—his successors would be more plausible promoters of his policies and less likely to provoke the necessary opposition to their execution.

History provides solace in this dark hour. Our worst moments have often been followed by remarkable recoveries—errant knavery replaced by wisdom; royal screwups yielding to rationality. It took a civil war, but we did pivot from cruel slave labor to giving black men the vote. The 1890s Gilded Age of extreme wealth inequality morphed, by the turn of the century, into the Progressive Era of better working conditions and regulation of outrageous business practices. The stock market crash of 1929 ushered in the New Deal and prosperity. A century of Jim Crow in the South made way, after little more than a decade of protest, to landmark civil rights legislation. And most recently, we careened from the military and economic disasters of the irrational George W. Bush to the relative peace and prosperity of rational Barack Obama.[1]

The scorecard of the last fifty years reveals a recurring pattern: failed Republican presidents, bailed out by more reality-based Democratic presidents. Nixon resigned his office when exposed as a liar and a crook. Reagan's "supply-side economics" created enormous deficits and his "freedom fighters" became today's terrorists. Baby Bush encouraged a financial bubble that burst into the biggest economic downturn since the Great Depression and started two expensive wars that show no signs of ending. Trump's lack of competence, ethics, and decency, in both person and policy, make it likely he will leave office either impeached or in a dogfight with

Baby Bush for the ignominious dishonor of being the worst president in United States history.

A more reasonable person and party will almost certainly take over, probably in 2020, with a mandate to clean up the Trump mess—just as Carter had to do after Nixon; Clinton after Reagan/Bush; and Obama after Bush II. And, sooner or later, the American people will catch on that the Republican (now effectively the Tea) Party has been selling pure snake oil—societal delusions that are bad for us now and will be even more disastrous for our children and grandchildren. As Abraham Lincoln put it: "You can fool all the people some of the time, and some of the people all the time, but you can't fool all the people all the time." The best thing about Trump, perhaps the only good thing, is that he is fool enough not to be able to fool very many people for very much longer. He is clearing the way for grown-up leadership.

Progressive Populism to the Rescue

> The Prince *was written by Machiavelli for the Haves*
> *on how to hold power.* Rules for Radicals *is written*
> *for the Have-Nots on how to take it away.*
> —SAUL ALINSKY

> *Those who love peace must learn to organize*
> *as effectively as those who love war.*
> —MARTIN LUTHER KING JR.

Political power in the United States now resides firmly in the hands of the "AstroTurf" populism so skillfully exploited by Donald Trump, the Koch brothers, and other radical-right extremist plutocrats. The Republican Party exists now in name only. It has been hijacked and absorbed by the extremist Tea Party, which borrows its name but no longer reflects traditional

Republican policies—no longer respects conservative caution, no longer provides checks and balances, no longer chooses patriotism over party, and no longer protects from attack the cherished institutions of American democracy. The Democratic Party, weak even in victory, is now scattered and ineffectual in defeat. The only antidote to current fake populism is future real populism. Cheap, misleading slogans must be countered with effective and meaningful programs that can win wide support and reverse the polarization that degrades our political processes.

Until Trump, I never appreciated the importance of populism or felt its spirit. In 1968, I lived less than a mile from Columbia University but didn't even bother to go uptown to check out the student demonstrations (much less participate in them). I was enjoying a great beach day when Martin Luther King gave his epochal "I Have a Dream" speech. I was MIA at anti–Vietnam War sit-ins despite hating the war and dreading my upcoming army service. When it comes to good citizenship, I have a lousy lifetime record. Never volunteered for a candidate or helped get out the vote. Never joined in a civil rights march or helped picketing strikers. Never participated in Earth Day.

My detachment has dissolved with age (getting less selfish and hedonic), with fear (of Trump's headlong attack on societal sanity), and with the conditional hope, and high statistical probability, that Trump can be stopped if people make the effort. Fortunately, Trump is also bringing out the potential populist in lots of other people. Massive demonstrations have been organized supporting women's rights, minority rights, gay rights, health insurance, truth, science, and the environment. With the executive, judiciary, and congressional branches all controlled by the party of Trump, the remaining hopes are the media and a bottom-up, citizen-by-citizen, authentic grassroots movement. And this requires all of us to jump in—even no-shows like myself. None of us can complacently sit back and hope someone else will come to the rescue.

Even though I pretty much sat out the 1960s, there were two inspiring grassroots leaders I couldn't help but admire, even if passively and from afar. Everyone with eyes to see, ears to hear, and a heart to feel was moved by Martin Luther King. He was to the twentieth century what Lincoln had been to the nineteenth and Washington to the eighteenth—the beacon of his nation and best expression of our better angels. The much less well known Saul Alinsky has had an even more pervasive influence on our current political life—but in a twisted way that would break his heart. His brilliant community organizing techniques, developed to help the little guy get a fair shake from the big guy, have been co-opted by the big guys to keep the little guy in his place. My first job, begun the year Alinsky died, was working in a community mental health program that was based on his principles. I experienced firsthand, on a daily basis, just how effective they can be when used by the right people for the right reasons.

Their personalities and strategies could not have been more different—Alinsky was in-your-face, abrasive, confrontational; King was inclusive and embracing. But both were uniquely gifted at winning with weak hands and defeating power with truth. Both were broadly inclusive in their populism. King began work as a crusader for black civil rights, but ended it as a crusader for all human rights. Alinsky started out in white communities, but spent much of his life working in black ones. Both defended the underdog and worked in the trenches. Both fearlessly faced personal and political dangers. Both were four-dimensional chess players, equally good at doping out short-term tactics and long-term strategy. Both were Moses-like, able to see the promised land from a distance, but not to reach it. And both won many battles, but neither won the war. Their shared tragedy was an inability to establish a broad-based coalition that could change the face of America and prevent the later takeover of fake Tea Party populism. King had the eloquence, moral stature, breadth, and celebrity—but his assassination robbed him of the time. Alinsky had great influence, during his life and since, but in more restricted situations, in smaller spheres, and his techniques were applied best by the powerful people he most detested.

Martin Luther King was the antithesis of Donald Trump. His decency, patience, and self-sacrifice are a mute rebuke to Trump's bigotry, impulsivity, and self-promotion. "Love is the only force capable of transforming an enemy into a friend. We never get rid of an enemy by meeting hate with hate." Sounds unrealistic, but it was effective, because King combined idealism with practical know-how, psychological understanding, and organizational skills. His basic tool was the massive nonviolent populist demonstration, played out on television screens across America—an ongoing, movable passion play that powerfully pricked America's dormant racial conscience. King seized the moral high ground early and never gave it up. White racist provocateurs couldn't shake his human dignity. Impatient and aggressive younger black activists never spurred him into adopting overtly provocative tactics. I am a skeptic, usually immune to hero worship, but MLK was something special. His sincerity cut through hypocrisy; his integrity exposed deceit; his goodness shamed badness; his clarity of expression vanquished political jargon.

King's populism was as lovingly nonconfrontational as populism can possibly be, without becoming ineffectual. He always turned the other cheek, by instinct, but also because it was effective strategy. "The nonviolent resister must often express his protest through noncooperation or boycotts, but he realizes that these are not ends themselves; they are merely means to awaken a sense of moral shame in the opponent. . . . The aftermath of nonviolence is the creation of the beloved community, while the aftermath of violence is tragic bitterness. Nonviolence does not seek to defeat or humiliate the opponent, but to win friendship and understanding." King fought the evil of racism, but never felt that racist people were irredeemably evil or beyond the pale of future alliance. Practitioners of racism were also its victims and he hoped to free them of it at the same time he was freeing their targets. Hate of one's enemies is self-destructive to the hater as well as self-destructive to the movement.

Before his premature (but not unanticipated) death, King was

trying to form the broadest of populist coalitions, aimed at re-
ducing all forms of economic, social, and gender injustice, and
ending the tragic war in Vietnam. A strong supporter of the labor
movement, he was shot in Memphis as he was about to participate
in a garbage workers' strike. Were he alive today, MLK would
be fighting against our societal delusions, for our collective wel-
fare, and for future generations. He would be marching for immi-
grants and sitting in with Occupy Wall Street; he would comfort
families of gun victims and shame the NRA; he would protect
foreclosed homeowners and exhort bankers to relieve their dis-
tress; and he would defend the environment against multinational
corporations. He would give voice to the people—not put self-
serving words in their mouth (Koch brothers, I think of you as I
write this).

King was especially dangerous to the elites because he could
unite otherwise divided causes within one populist tent. There
was synergistic power in the numbers and in the shared commit-
ment. His assassination was a turning point in our history because
it cut short this second and even more important chapter of his
life's work. His moral force extended to all areas of American in-
justice, and his practical skills in mobilizing millions might have
led our country in a very different, and much more productive,
direction. His sane and humane populism might have prevailed
against the fake, retrogressive Tea Party populism that rules us
today. Trump might not have won the presidency had King been
blessed with normal longevity. A world influenced by his inclu-
sive, loving populism would have no room for Trump's hate and
exploitations. We can't wait for another MLK—he was a once-in-
a-century phenomenon—but we can be inspired by his memory
and learn from his techniques.

Saul Alinsky published his legacy book, *Rules for Radicals,* in
1971, shortly before his death. A manual for community organiz-
ers, it contains ten chapters crystallizing his thirty years of com-
munity organizing—detailing how to change the world, bottom
up, one small step at a time. Alinsky's genius was in helping people

determine their own fate. He had clear and compelling advice on the necessary precondition: "No matter how imaginative your tactics, how shrewd your strategy, you're doomed before you even start if you don't win the trust and respect of the people; and the only way to get that is for you to trust and respect them."[2]

His approach to community enablement was nonviolent, but highly confrontational and, in many ways, opposite to King's. Alinsky brought a community together with color-war tribalism, emphasizing similarities among its members and sharp differences from opponents—while King sought to find commonalities. Alinsky probed for ways to provoke conflicts that would enhance community solidarity via hostility toward a common enemy—while King discouraged provocation and sought ways to reduce conflict. Alinsky sought to beat opponents—while King wanted to join them. Alinsky needed a villain to coalesce community consciousness—while, for King, there were no villains, just people who were temporarily misguided and may later become friends. Both staged highly publicized nonviolent demonstrations and both exploited violent overreactions. But while the purpose for King was to shame the opposition into goodness, for Alinsky it was to humiliate them into surrender.

Alinsky's *Rules for Radicals* reads just like Machiavelli, only the advice he gives was meant to help the common man, not the Prince. Here's a summary of his recommendations:

1. You have as much power as your opponent thinks you have.

2. People power can fight money power.

3. Keep within your expertise and push your opponent outside his.

4. Ridicule can diminish your enemy.

5. Tactics that are fun to do are more likely to be followed and to work.

6. Keep the pressure on.

7. Stay one step ahead—as the opponent figures out defenses, change tactics.

8. Violence on the other side wins you friends.

9. Pick a target and personalize it.

10. People fold faster than institutions.

Alinsky devoted his life to the righteous cause of helping the powerless defend against the depredations of the powerful. The tragic irony is that his techniques have since been systematically copied by the powerful to gain even more control over the powerless. The Koch brothers' organization of the Tea Party followed his playbook, turning Alinsky techniques into weapons to subvert Alinsky goals. King's nonviolent populism was based on an overriding morality. Alinsky's techniques were practical, tactical, utilitarian, and equally usable by both sides in every conflict—as much sauce for goose as for gander. Alinsky hoped to fight money power with people power. The Kochs cleverly used money power to buy people power and the resulting Tea Party "populism" has been successful in dragging America to the extreme right.

When King was killed in that awful moment of racial hate fifty years ago, no one would have predicted that a black man would so soon be elected to the presidency of the United States. This was a redeeming moment in American history and seemed to offer a great opportunity for further populist progress—but it turned out to be an opportunity mostly lost. Obama won reelection and remains personally popular, but during his eight-year watch, extremist elements of the Republican Party gained hegemonic control of Congress and of most state governments. They did this with discredited policies and, in many instances, with less-than-credible people. The irony was that fake populism could so easily overwhelm the real populism of a president who prided himself on skills as a community organizer. Of course, the Tea Party had advantages on its side—very big money, shameless cynicism, un-distracted pursuit of thinly disguised self-interest, Fox News, and talk radio.

Obama's first job after law school was as a community organizer in Chicago; he was elected president in part through the efforts of a populist Internet community (MoveOn.org), and he may well serve his postpresidency years as a community organizer on a very much larger scale. But to his own and the country's great misfortune, Obama stopped being a community organizer and populist leader once he got to the White House. He let himself get stuck in the quicksand of "inside the Beltway" politics—in the death grip of a stubborn, Tea Party–dominated Republican leadership, willing to let the country sink so long as Obama sank with it. Obama might have accomplished much more had he spent less time, labor, and political capital trying to find common ground with the intransigent Republican Party hacks in Congress and instead spent more effort going above their heads directly to the American people.

By strange coincidence, most of Alinsky's work had taken place in Chicago, in the same neighborhoods where Michelle Obama grew up and where later Barack Obama had his first job. And to add an extra layer of coincidence, Hillary Clinton was once offered a job by Alinsky and wrote her college thesis on his work. Obama would have done well to take his lessons from the confrontational Alinsky playbook that was guiding his Tea Party opponents. But by temperament, Obama is much more like Martin Luther King, eager to compromise, even with people who made it abundantly clear that they would never compromise with him.

For reasons that still bewilder me, Obama didn't appear to understand the importance of engaging the people rather than fighting the politicians. When I first heard Obama speak to the 2004 Democratic National Convention, I was moved in a way that I hadn't been since King. He was the most articulate speaker

of his time, but only in isolated speeches. He never created a pow-
erful, direct alliance with the American people as FDR had with
his "fireside chats." He should have fought his Republican op-
ponents less on their stonewalling terms and more on the public
stage, exposing their disingenuousness and self-serving hypocrisy.
His charisma might have elicited a more coordinated, public-
spirited approach to our problems—in place of constant sparring
over slices and crumbs of the pie. Stuck within an interminable,
soul-destroying battle with the petty Republican politicians of
Congress, Obama failed to rise above them and to raise us above
the sectarian squabbles.

When people know they're selling lies, they work doubly hard
to perfect their salesmanship; people selling truth complacently
and mistakenly assume it will sell itself. If you consciously set out
to deceive voters, you work hard to understand their psychology
and tailor your language so that it delivers a maximally persuasive
message. The Koch brothers and like-minded extremists have
spent thirty years and tens of billions of dollars building a network
of think tanks to clothe their wolfish agenda in common-man
sheep's clothing. A vocabulary of cleverly concocted buzzwords
disguises selfish motives under pious hypocrisies. Unchallenged
lies and innuendoes, constantly repeated, become accepted wis-
dom, invulnerable to contradictory facts.

Once implanted, unconscious cognitive biases become the lens
through which we view the world—and they are extremely dif-
ficult to change, especially when also shared by your group.[3] The
Tea Party could absorb the Republican Party, and then vanquish
the Democratic, because it was smarter, better funded, more psy-
chologically minded, more focused, more cynical, and much more
ruthless. Its special skill is on attack, twisting language to turn
sins into virtues and strengths into liabilities, with a compelling
terminology derived from Orwellian Newspeak. Joining a "Tax
Revolt" means unwittingly supporting tax evasion by the super-
rich and the multinational corporations. "Getting Government
off Our Back" means freeing big banks, Big Pharma, and Big

Energy to have their way with unwary consumers—unregulated by government protections. "Small Government" means worse schools, health care, and public services for the ordinary people who can't afford to go private. Support for "Family Values" means defunding programs that might actually help keep families together. Protecting a fetus's "Right to Life" allows you to not worry about its quality of life once born. "Welfare Mothers" are irresponsible and should be kicked off the government dole, but Big Oil and Big Agriculture get to keep their big government subsidies. "Entitlements" are evil unless it is the military contractors who are entitled. "Redistribution" of wealth is un-American unless the wealth is flowing upward to the superrich. "Individual Liberty" and "Constitutional Rights" should prevent government interference with gun owners, but should not prevent government interference with a woman's womb. "States' Rights" must be protected, unless it is a blue state exercising its rights against the right-wing agenda. Good manners and decency are deplored as "Politically Correct"; bad manners and interpersonal cruelty are extolled as "telling it like it is." "Class War" is not the attack of the superrich against the lower classes, it is any attempt by the lower classes to mount any form of defense against the superrich.

In classic Alinsky style, the Tea Party demonizes its enemies, consolidates its constituents, and seeks converts. Its complete lack of idealism makes it much better at backroom deals than progressive populists—who coddle their enemies, allow dissension within their ranks, and preach to the choir.[4] Various wings of the left fight among themselves over ideologically pure distinctions that don't make any real-world difference in the face of the Tea Party takeover. And Dems aren't cynical enough to win by compromising principle for expedient result, a fatal flaw in our current cynical political civil war. Right-wing coalition builders have fewer scruples about co-opting potential opponents by giving them a piece of the action. In the past, progressives like Abraham Lincoln and Lyndon Johnson were great at this kind of manipulative wheeling and dealing—it was essential to their success in pushing civil rights bills through reluctant congresses. Horse trading was beneath Obama's dignity, and he (and we) have suffered for it. The Tea Party fought dirty, while Obama fought with one hand voluntarily tied behind his back.

Patronage politics creates strange bedfellows and honor usually dies where self-interest lies. Trump has a hard-core base consisting of evangelicals, alienated rural residents, tax dodgers, corporate deregulators, gunslingers, white supremacists, anti-Semites, and conspiracy theorists. Progressives tend to be more finicky in their choice of allies and more self-destructively idealistic. In blogs, in tweets, and in person, I tried, but failed, to convince Bernie voters that blocking Trump should be their highest priority—Hillary might not be perfect, but Trump was perfectly awful. Many disagreed and insisted on voting Green Party, saying noble somethings like "I must always follow my conscience." It reminded me that Ralph Nader never apologized for saddling us with George Bush, doubtless because he too could claim he was just following his conscience. The preciously independent and divisive consciences of the left may be its undoing, and ours.

The extreme right speaks piously while simultaneously dealing cards from the bottom of the deck. Its hypocrisy hasn't fooled, or endeared it to, the pope. He deplores people who defend Christian values for political purposes, while failing to live by the essential Christian virtues—and openly expresses a preference for atheists who live a moral life over religious bigots who spout morality but fail to live by it (most evangelical leaders, this means you). The pope does not see a clear path to heaven for those whose lives are governed by the blind pursuit of wealth, indifference to the poor, conspicuous consumption, and heartlessness toward the suffering. Does anyone on God's earth think Jesus Christ could ever, under any circumstance, bring himself to vote for Donald Trump? How impossible to picture Donald Trump fitting through the eye of heaven's needle.

Tea Party success in cognitive framing results in a stunning disconnect between what our people want and what we are getting.

A consensus of Americans support positions that are middle of the road, equivalent to today's Democratic platform (and yesterday's Republican). But we just elected its polar opposite in Trump and have given his Tea Party types control of all branches of the federal government and most of the state governments. What a sad reflection on the pathetic political skills and disunity of the Democrats and what a scary testimony to the political guile and lockstep march of the Tea Party Republicans.

How do "we the people" get back in the populist game? How do we retake our government so that it expresses our views, not the interests of the 1 percent? First off, by mobilizing "we the people." Throughout our history, Americans have made enormous sacrifices to fight the external enemies of democracy. We must now rise to this occasion when the threat is internal and, in many ways, more insidiously dangerous. Trump and his cronies will keep pushing until there is sufficiently strong pushback. Our responsibility for protecting democracy should become second nature—like putting on seat belts. We don't think twice about setting aside time and attention each week for pursuits like PTA or church worship. For progressive populism to work, we must set aside time and devote effort to "nurture and protect" democracy by participating in peaceful demonstrations, contacting members of Congress, writing and signing petitions, raising money for progressive organizations, and hosting or attending citizen coffee klatches. I never thought I would be saying this, but late in life, and courtesy of the Trump threat, I have finally seen the light.

The Internet is the tool of modern populism, having already mobilized powerful forces in every country except those that have strict censorship (think Arab Spring, the Ukrainian Orange Revolution, Brexit). Formal elections occur years apart, while Internet influence can be exerted literally in seconds and with tiny transaction costs. Obama was our first Internet president. The Tea Party was our first Internet political movement. After conquering the Republican Party, it thwarted Obama and diminished his legacy, and now it runs the country almost unopposed. The Internet will be the weapon of choice for both sides in every future political struggle. Our Tweeter-in-Chief is sitting in the Oval Office precisely because he understood that 140 characters were much more important in swinging the collective mood than detailed policy statements.

Progressive populism has its own effective Internet weapons—although so far they have lost the social networking battle with the Tea Party. The oldest and largest is MoveOn.org, started casually and with no budget in 1998, by two Silicon Valley types. They posted a petition to Congress to "censure President Clinton and move on" (from the salacious Monica Lewinsky scandal). When it went viral, they knew they were onto something. Using their initial list of signers, and their well-honed tech skills, the pair launched what soon became a multifaceted, well-organized online juggernaut aimed at defending the 99 percent Davids against the 1 percent Goliath. When the Democratic Party dodged its responsibility to buck Bush, MoveOn became the default leader of opposition to the Iraq War. It was key to Obama's election and reelection and has been on the right side of every political question for the past twenty years. Its technical campaigning and research skills are highly developed and it keeps close to its grass roots. MoveOn is admirable and very good at what it does, but it doesn't do enough to be a meaningful counterweight to the Tea Party. In part, this is a function of funding (average contribution: twenty dollars); partly its lack of talk radio and other media backup; partly its unwillingness to participate in dirty political tricks; and mostly it's because MoveOn's grassroots support is too narrowly focused on urban, intellectual elites. Success for the 99 percent against the 1 percent requires a unified 99 percent, and MoveOn has so far failed to achieve anything approaching unification.[5]

Indivisible, another great online populist resource, was launched by former congressional staffers frightened by the Trump/Tea Party takeover. They compiled an "insider's" guide, which went viral, instructing ordinary people how to apply pressure on their representatives to force them to become truly representative. It is a remarkably rich, one-stop compendium of everything you need to know to become part of the political process; has already been downloaded millions of times; and supports more than five thousand local groups formed to resist Trump and his agenda. The early results are promising. Politicians who might have hidden in the Trump weeds have been called out at town halls, shamed on the media, and flooded by calls and messages. Some seem to be showing surprising spine in opposing the Trump administration on Obamacare, his Russian election manipulations, and his choice of cabinet appointees.[6]

SwingLeft is another exciting groundswell movement focusing its efforts on winning back the House of Representatives in 2018—restoring checks and balances to a government that is now as badly unbalanced as our government has ever been. The goal seems almost impossible to accomplish—because of gerrymandering, only sixty-five congressional districts remain competitive and Democrats would have to win 65 percent of these. SwingLeft's motto, "Don't despair, mobilize," is not completely unrealistic because midterm elections usually favor the party out of power and because of Trump's strange behavior, unpopular policies, and incompetent management. Their website is simple and user friendly. You click an icon to discover which is your nearest swing district and how to join/host a "house party" launching a bottom-up, community-canvassing effort to identify your neighbors' concerns, attitudes, and preferred solutions. The most important outreach is to swing voters and people who might not otherwise vote. The approach is to listen, not preach, to find common ground, and to fight Tea Party money power with people power. SwingLeft recently launched with five hundred "house parties" around the country and hopefully will go viral. I've met some early adopters and think that, with our help, they may have a shot at starting a contagion.[7]

OurStates is designed to influence those state legislatures that are moving very fast to implement the Trump/Koch agenda on economic justice, immigration, reproductive rights, voting rights, gay equality, and policing. The site provides detailed information on which states have important pending legislation and how to influence the process in a positive way. It instructs people on how to find their state legislators, meet with them, frame the "ask," and follow up. There is also useful and practical advice on how to pressure unreceptive state legislators to make them work for the public, instead of their campaign contributors. The most crucial efforts at the state level will be to win back sufficient seats before 2020 to reverse the inequities of GOP gerrymandering and restore democracy to the states.

Many other populist groups have formed, and doubtless dozens more will form in the future. Some of the existing populist initiatives are general, some are issue-oriented (travel bans, immigration, women's rights, science denial, Supreme Court nominees, Trump's psychological health, etc.). The lack of the usual governmental checks and balances places the heavy responsibility of restraining Trump on "we the people" and our (so far) free press. Opposition to Trump will mount whenever he grabs for power and/or screws up—both of which are likely to be frequent events.

Trump's base may be much less solid than it appears, especially since he is already betraying it by his attempts to rob many of them of health insurance; by his Russian connections; by his appointing a cabinet of billionaires; and by his suggesting tax cuts that focus on benefiting the superrich. Disillusionment provides opportunity for a whole succession of anti-Trump populist petitions and demonstrations that eat away at his base: "Trump Voters Against Trumpcare"; "Trump Voters Against Trump Tax Cuts for the Rich"; "Vets Against Trump"; "Evangelicals Against Trump"; "White Women Against Trump"; "Gun Lovers Against Trump"; "Rural America Against Trump"; and even perhaps a "Putin, Go Home" movement for those disgusted by the Trump/Putin manipulation of our democratic election processes. Trump's approval ratings during his early tenure in office are a shocking 20 percent lower than the average of previous presidents. He is a decidedly unpopular minority president, with no mandate from the people, wrenching our government in a radical direction, oblivious to its many risks. Trump persists in a crazy claim that he would have won the popular vote by two million (instead of losing by three million) had there not been extensive voter fraud allowing five million flawed ballots to be cast for Clinton. There is not a scintilla of evidence of even one fraudulent vote for Clinton. Trump's lie is rejected by everyone (Republican and Democrat alike), except for Trump's protective inner circle of family and sycophants. As his base erodes, opportunist politicians will jump off his bandwagon and begin to reel him in.

But we mustn't ignore a serious voter fraud scandal, skillfully orchestrated by the Republican Party and funded by the Koch brothers, that does besmirch our democracy. When the GOP-dominated Supreme Court failed to continue its long-standing support for the Voting Rights Act, the GOP-controlled legislatures in twenty-two states jumped in with new voter restriction laws (fourteen of which were enacted just in time for the 2016 election). In Wisconsin, which Clinton lost by only twenty-seven thousand votes, three hundred thousand people, mostly Democrats, were prevented from voting due to photo ID requirements. Progressive populism must carry the fight for reenfranchisement of those who have been improperly disenfranchised, as well as doing the on-the-ground work of getting voters to the polls. There is also an enormous opportunity to regain the loyalty of people who voted for Obama, but not Clinton, either because they stayed at home, crossed over to Trump, or voted Green.[8]

The Tea Party controls our government despite representing the views of no more than a third of the electorate—it's time for the silent majority to take back our country. There have already been repeated mass demonstrations against Trump in most U.S. cities and in cities around the world. Alinsky and King disagreed in many ways but converged on the absolute necessity that the progressive movement must have zero tolerance for violence and must assiduously weed out the bad apples. Every demonstration is a competition—you lose if you throw the first stone, you win if your opponent does. Self-discipline, even in the face of the most extreme provocation, wins the moral high ground and deprives Trump of his ready-made "national security" excuse for grabbing more power.

Nonviolence doesn't mean nonconfrontation. We need to channel Alinsky when dealing with unbending opponents (think Tea Party hacks in Congress). Ignoring small lies leads to bigger and bigger lies. Every "alternative fact" must be checked and repudiated. Every cynical piece of proposed Trump legislation must be opposed vigorously, every appointment challenged. Cronyism, corruption, and conflicts of interest exposed. Trump's cabinet is mostly filled with people who are morally and/or intellectually poorly equipped for their leadership positions. Requisite congressional oversight will not occur unless an outraged public persistently demands it. Setting low expectations on Trump's presidential behavior will enable him in even more extravagant displays of impulsivity and arrogance. He must be called out early and often. The most important populist goal, at least until the 2018 election, is to force and cajole Congress into reining in Trump. The populist movement must stiffen the spine, and force party unity, of the often spineless and divided Democrats. It needn't waste efforts with those Republican congressmen who are sincere Trump lovers or with the few who have already been patriotic enough to stand up to Trump. Instead, populism must focus on the majority of Republicans who are cynical opportunists—using Trump to pass their donors' extremist agenda, while holding their noses and hoping to tame his worst excesses. At least some may become more afraid of their broader constituencies than they are of Trump, their donors, and the Tea Party. And most important, the people must defend freedom of the press. Trump's "the press is the enemy of the American people" must be vigorously opposed with Jefferson's "our liberty cannot be guarded but by the freedom of the press, nor that be limited without danger of losing it." The fight has just begun—it will be a war of attrition, winnable only if the pressure is significant and persistent.

Jesus Would Tell His Flock to Vote Righteously, Not Radical Right

Trump won the evangelical vote by a surprisingly wide four-to-one margin and won the white Catholic vote by two-to-one. This was no tribute to Trump's religious purity—rather, it was the work of cynical Christian leaders who had sold their souls to Trump in a shady backroom deal. They would influence tens of millions of religious voters to support him in exchange for his support for their hard-line positions against abortion and gay rights. The wheeling and dealing was remarkable testimony to the political skills, as well as the religious hypocrisy, of most of the Christian leaders in the United States. Thirty pieces of silver never exchanged hands, but the teachings of Jesus Christ were surely cast by the wayside.

Jesus didn't care a fig about abortion or homosexuality. In his time, abortion was legal and widely practiced—but he never once condemned it in all his many preachings. Homosexuality was also accepted and widely practiced—and again Jesus never once condemned it. Jesus was the champion of the underdogs against fat cats like Trump. Christ honored the humble and the weak: "Blessed are the poor in spirit, for theirs is the kingdom of heaven. Blessed are the meek, for they will inherit the earth." "For it is the one who is least among you all who is the greatest." He would never horse-trade the needs of the poor and the oppressed to further a "fundamentalist" religious agenda and billionaire-inspired right-wing causes.

Anyone who believes that Jesus could ever support a man like

Trump needs lots more Bible study. Jesus was a world-class giver. "Give to everyone who begs from you; and of him who takes away your goods do not ask them again. And as you wish that men would do to you, do so to them." "If you want to be perfect, go, sell your possessions and give to the poor, and you will have treasure in heaven." "When you give a banquet, invite the poor, the crippled, the lame, the blind, and you will be blessed. Although they cannot repay you, you will be repaid at the resurrection of the righteous." In contrast, Trump is a world-class taker—he invites the billionaires to feast on the poor.

And how does Trump measure up to Christ? "You shall not commit adultery, you shall not kill, you shall not steal, you shall not covet are summed up in this single command: You must love your neighbor as yourself." Trump is a serial adulterer, a business thief, a tax cheat, and a greedy coveter of epic proportions. He brags about being above the law of both God and man in these most remarkable words: "I could stand in the middle of Fifth Avenue and shoot somebody and I wouldn't lose voters." Trump hasn't directly killed anyone, but his attempt to deprive health care could kill millions and his promotion of global warming may wind up killing tens, or even hundreds, of millions.

Jesus was one of history's most forgiving people—but he could not tolerate religious hypocrisy. A sampler from among his many denunciations: "And when you pray, you must not be like the hypocrites." "You hypocrite, first take the log out of your own eye, and then you will see clearly to take the speck out of your brother's eye." "You hypocrites! This people honor me with their lips, but their heart is far from me." Jesus could accept abortion and he could accept homosexuality, but he could not accept the hypocrisy and lack of charity so baldly displayed by the radical religious right. Faced with Jerusalem versions of Trump, Jesus unceremoniously kicked them out of the temple. He declared that passage to heaven would be as difficult for a rich man as a camel going through the eye of a needle.

Pope Francis lives according to Jesus's teachings on the poor and also shares his disgust with religious hypocrisy: "There are those who say 'I am very Catholic, I always go to Mass' . . . but they don't pay their employees proper salaries, they exploit people, do dirty business, launder money. . . . To be a Catholic like that, it's better to be an atheist." His message to religious hypocrites: "You will arrive in heaven and you will knock at the gate: 'Here I am, Lord!'—'But don't you remember? I went to Church, I was close to you . . . don't you remember all the offerings I made?' 'Yes, I remember. The offerings, all dirty, all stolen from the poor. I don't know you.' That will be Jesus's response to these scandalous people who live a double life." Asked about gays, Pope Francis replied: "Who am I to judge?" The pope washes the feet of the poor; Trump tries to steal their health care, so he can deliver yet another tax giveaway to the richest 1 percent.

By becoming political heavy hitters, many Christian leaders in the United States defy the most basic tenets of Christ's Christianity. They ignore his warning: "Beware of false prophets, who come to you in sheep's clothing but inwardly are ravenous wolves." They ally themselves closely with the Republican Party (the ravenous wolves Jesus was talking about) in a partnership to exploit the poor and further enrich the very rich. "For what shall it profit a man, if he gain the whole world, and suffer the loss of his soul?" True believers in Jesus must want health insurance for everyone, housing for the poor, treatment for the mentally and physically ill, adequate care for mothers and children, protection for persecuted immigrants, and stewardship of the earth. It is hard to conceive of a human being less Christian in word or behavior than Donald Trump. Many responsible Christians have chosen to follow their consciences and buck their leadership (often to their own detriment), speaking out against Trump's policies and person. People who take Christian values seriously should follow their lead and form their own political judgments, not follow the strictures of hypocritical evangelical leaders who spout, rather than live, the Bible.[9]

Leading the Populist Charge

Ever since Thucydides, 3,400 years ago, there has been a run-ning controversy among historians over whether it is great lead-ers who make history or whether history makes great leaders. Do leaders appear great only because they are surfing the wave of deeper geographical, demographic, climatic, and social forces, or are they able to exert a powerful and independent driving force? The Trump experience raises the equally interesting antipodal question that rarely gets asked—how much does a really dreadful leader change the flow of history, or does he merely express it? I come down on both sides of these two questions, placing my bet on historical forces for the long run, on great (and dreadful) leaders for the short. Trump has shown us how much damage a terrible leader can do. It would be nice to see (as soon as possible) how much damage repair can be accomplished by a great one. It remains an open question whether fresh, competent leadership can move us toward the commonsense goals laid out later in this chapter in the "We the People Contract." Perhaps even the wisest and most visionary leaders will inevitably fail in the face of such severe problems, the power of well-funded political opposition, the inbuilt flaws of human psychology, and the increasing fragility of our planet. Certainly, Obama was both wise and visionary. Was his failure inherent or contingent? Can someone else successfully push the reasonable agenda his opponents so ably scuttled?

History provides us with hopeful political precedents—transformative presidents who changed our attitudes and set a new course for our nation's direction. The United States might never have existed at all, and certainly would not have existed in its current form, were it not for the great leadership of George Washington. Three score and ten years later, we could not have preserved our Union without the great leadership of Abraham Lincoln. And a further seventy years on, the Great Depression would have been lots greater without the interventions of Frank-

lin Delano Roosevelt. When FDR took office, the United States
was facing far worse problems than ours today and required far
greater institutional changes than would be necessary now. Mar-
kets were broken throughout the world, trade was at a standstill,
a quarter of workers were unemployed, people were starving, and
the economy and everyone in it seemed paralyzed. But none of
this discouraged FDR. He felt confident and knew how to inspire
confidence in others. His first message resonated throughout the
country: "There is nothing to fear but fear itself." And he con-
tinued to connect on a deep emotional level with the American
people through his "fireside chats." Roosevelt understood that
to cure our nation's economic depression, he had first to cure its
psychological depression. Herbert Hoover was a very competent
man, but he couldn't cure our depressions—partly because his
policies were wrong, but even more because he didn't have Roo-
sevelt's personality and interpersonal skills.

Yet another seventy years later, many of us hoped Obama might
also become a transformative president. His campaign had captured
the FDR spirit with his "Audacity of Hope" and "Yes, We Can."
He was among the smartest, most logical, truthful, even-handed,
objective, and least egotistical of all U.S. presidents. Electing an
African-American to our highest office was inspiring evidence of
racial progress since Lincoln's time. And most important, Obama
had a deep understanding of all the dangers posed by every so-
cietal delusion and offered the most rational proposals to avoid
them. Our country, and our world, would be much further along
toward sustainability had we followed his vision and enacted
his policies. Obama had inherited a simply terrible hand from
Bush—an economy in free fall and two unwinnable wars. He
played the bad cards he was dealt cautiously and well—restoring
the economy and avoiding, for the most part, further foreign
policy blunders. But having entered office with the potential for
greatness, he left it a mostly failed (and certainly not a transfor-
mative) president, rendered almost powerless by a cynical and
stubbornly uncompromising Republican opposition whose stated

goal was to destroy him, even if that meant harming our nation. But Obama also suffered from limitations imposed by his own personality—he was, quite simply, too normal and too nice. His weaknesses as president came from his strengths as a man—his humility, ability to see both sides of every question, cool reserve, and dispassionate objectivity.

The most transformative presidents—for good and for ill— have often been among the most narcissistic—not normal, nice guys. Their ability to bend, or at least postpone, the course of history depends on having an outsize ego, seemingly unrealistic ambitions, and the gift of engaging, at some very deep level, the gut emotions of the public. By simple force of personality, they can move mountains and change what previously appeared to be the rules of the game. Whether you agree or disagree with their policies, Roosevelt, Reagan, and Clinton all clearly transformed the politics of the last century. Unfortunately, Obama could not do the same in this one. And, equally unfortunately, Trump can. The worst presidents have also been among the most narcissistic, and Trump has, in his own weird and destructive way, already managed to become one of the most transformative presidents in our history. We should expect our leaders to be narcissistic, and maybe not even to be the nicest of guys (or gals)—the question is whether they are using their narcissism on behalf of, or against, the people.[10]

Effective populism is never leaderless because populism without leadership invariably devolves into anarchy. But no leader can bring us back to sanity unless he or she can harness the power of a progressive populism, while containing the passion stirred by fake populism. We are basically pack animals inclined to follow the pack whether it runs in the best or the worst of directions. Our opinions on issues are fickle and very situation- and leader-specific. The right leader, giving the right message in the right way, can move us in the right direction. The wrong leader, powerfully delivering the wrong message, can lead us to disaster. Trump has smothered the audacious Obama "Yes, We Can" hope with a dismally resounding "No, We Can't" dysphoria. But Trump is a dead end and hopefully a temporary aberration. We must hope that our next generation of leaders can move the people back toward facing reality. America has always been about second chances. We need one now.

To pull us out of the Trump dark age, a leader must have the right combination of policies, personal charisma, empathy, communication skills, and genuine ties to the people. An AntiTrump must be the spokesperson for real populism, not the fake storebought kind. In the last election, Bernie Sanders came closest. My heart was with him, but my head incorrectly thought that Clinton was the much safer bet to stop Trump. Bernie was a noble romantic, terrific on policy, and connected beautifully with his followers. But Bernie was probably too exotic a bird to be the AntiTrump transformative leader we need. He was the perfect populist for the intellectual crowd, but not for the country crowd—too city, too old, too Jewish, too preachy, too easily tarred with the dread buzzwords "socialist," "liberal," and "revolutionary."

We must find a more crossover AntiTrump president, someone who can bring out our better angels on both sides of the polar gaps and who can replace dividing buzzwords with commonsense solutions. Bill Clinton, with all his many policy and personal faults, is the prototype for a future AntiTrump. He was a broad-spectrum progressive populist, whose appeal ranged from eggheads to Bubbas; he was equally at home on spiffy college campuses and at county clambakes, among blacks and whites, with billionaires and poor folk. It was no accident that he was the last Democrat to win in rural America. Clinton had unusual persuasive abilities, conveying empathy and framing complex issues in simple, readily understandable terms. He could get people to accept sacrifice and do the right thing. Only he could have gotten Americans to forgo the short-term joy of tax cuts so we could pay off the national debt—and thus indirectly to protect future Social Security benefits. Clinton's personal conduct was surely a dishonorable blot on the office of the president, but that didn't stop him from being a very effective leader and bringing the people together. We won't have a transformative, people's president in 2020 unless opponents to Trump begin to unify now on the issues and the leaders who best represent them.

What Politicians Can Learn from Psychotherapy

I spent forty years of my life immersed in psychotherapy—as trainee, practitioner, patient, teacher, supervisor, researcher, and reviewer of research grants. What I learned smoothed off many of my personal rough edges and still comes in handy during life's most difficult moments—when I am consoling a dying friend, soothing a lost soul, resolving a spat with my wife, negotiating a contested bill with my repairman, containing my wiseass grandkids, or advising a colleague on an important life decision. Psychotherapy isn't something you just do. It is something that also becomes you.

Psychotherapists and politicians have a great deal in common, sharing very similar goals and techniques, even though differing widely in their scope of influence. Both change attitudes and behaviors by understanding and appealing to spoken and unspoken motivations. Psychotherapists work one patient at a time, while politicians impact millions, but the skill sets are alike. I learned a lot studying the moves of effective politicians, and I think politicians can become more effective by studying the moves of psychotherapy. Therapeutic cunning will be particularly necessary for politicians working to cure the American insanity Trump embodies and exacerbates.

Only rarely will a psychotherapist directly confront a delusional patient with fact-based arguments aimed at proving that his beliefs are false and self-destructive. However ridiculous and impairing the delusions may seem to the outside observer, they have helped the patient compensate for a painful reality—and won't be given up just because they are wrong and harmful. Premature attempts to impose reality result in the patient feeling anger, anxiety, and bewilderment. He is likely to become even more stubbornly delusional and unwilling to work with you. The truth may sometimes set you free, but you must be ready to hear it— and it must be delivered at just the right time and in just the right way. When you see a delusional patient in the emergency room, you must first gain his trust as a precondition for exploring the fears, feelings, fancies, stressors, legitimate beefs, and experiences that have made the false beliefs so believable. A good psychotherapist validates the underlying distress that necessitates a delusion, while at the same time working gradually with the patient to find more realistic ways of alleviating it. He expresses empathy for the patient's suffering, without validating the patient's delusional avoidance of its underlying causes.

Politicians will need a similar strategy to slowly and gently bring our society back to reality. Societal delusions are serving a perversely useful purpose for those who promote and believe them—and won't be given up just because they are wrong and dangerous to our world. Politicians lose elections when they bluntly ask voters to bravely face facts (think of Jimmy Carter's "malaise" speech).[11] In contrast, Trump's secret weapon in winning was his unique ability to nurture societal delusions for political and personal gain. He ruthlessly exploited genuine fears and insecurities, validated long-held grievances, and offered himself as the Second Coming—using patently false promises that persuaded otherwise sensible people to believe in him, even when he was telling the most obvious and outrageous lies. The Trumps of the world are immune to ever having any close relationship to the truth, but many of his voters would have been less susceptible to his con game if approached by an opponent possessing greater political and psychotherapy skills. We can't count on the timely appearance of another Abraham Lincoln or Franklin Delano Roosevelt. But the psychotherapy playbook does provide useful hints on how best to appeal to voters' better angels in countering Trump's exploitation of their inner demons.

One of the first and most important tasks of the psychotherapist is to place himself in the client's shoes—to start with the premise, "If I were in this person's circumstances, then I might act, think, and feel as he does." However different in smaller ways, we are all basically human in larger ones, sharing similar needs, fears, and frustrations and responding in similar ways to life's exigencies. It doesn't require a giant leap to imagine what it feels like to have your livelihood permanently wiped out; to be neglected, misunderstood, and misrepresented by politicians; to live under a government that seems all-encompassing, but unresponsive to your needs and fears; to be disdained by the media and the intellectual elites; to be standing still in line, while everyone else (for example, women, minorities, and immigrants) seems to be cutting ahead; and to practice traditions that many others no longer respect. Being one down spurs anger toward those who are one up. Resenting minorities is easy when you feel they are usurping your prerogatives, competing for your jobs, stealing your benefits, and marrying your daughters. Religious orthodoxy is a life preserver in a rapidly changing and alienating world. Discounting science allows the comfort of illusions. And add to all this findings from cognitive and neuroscience that people with conservative political leanings are psychologically and biologically inclined to respond more strongly to fearful situations.

Trump could not have won if his base were restricted to bankers and businessmen. He won with the enthusiastic support of ordinary people—whose interests are far better served by Democrats, but whose psychology is far better understood (and exploited) by Republicans. Democrats have lost their natural constituency because they stopped taking the trouble to understand it. Republicans gained an unnatural constituency by being much better psychologists and much slicker salesmen. To win and govern effectively in the future, Democrats must learn to be better psychologists and better salesmen.

The essentials of a therapeutic alliance in psychotherapy are equivalent to the essentials of an effective political alliance. Here are some of the cardinal rules:

- Be genuine and encourage genuineness.

- You can't help patients unless you form a strong tie with them.

- Speak the patient's language.

- Listen carefully and learn as much from your patients as they learn from you.

- Let patients know that all your efforts are focused on them.

- Empathy and trust are the most essential ingredients.

- Encourage patients to freely express their pain, fear, anger, disappointment.

- Determine their needs and how they want them filled.

- Discuss realistic goals and expectations.

- Don't be judgmental.

- Instill hope.

- Metaphors, images, and parables are more effective than facts and figures.

- Be aware of your own feelings and use them effectively.

- Everything doesn't equal everything else—less than 10 percent of what's said in psychotherapy accounts for more than 90 percent of change. Always be alert to, and do everything you can to leverage, the potential tipping points.

Wherever you see "patient" above, substitute "voter." Great politicians are born, not made. They know all this instinctively by the time they are elected president of their middle school class. But good politicians can become much better by learning the lessons of psychotherapy and applying them in their day-to-day work with constituents.

António Guterres, secretary-general of the United Nations, learned from his psychoanalyst wife the political value of psychological insights: "She taught me something that was extremely useful for all my political activities. When two people are together, they are not two but six. What each one is, what each one thinks he or she is, and what each one thinks the other is. And what is true for people is also true for countries and organizations. One of the roles of the secretary-general when dealing with the different key actors in each scenario is to bring these six into two. That the misunderstandings disappear and the false perceptions disappear. Perceptions are essential in politics. And in it is not just bringing six into two; the task is often coordinating hundreds so that they can act as one to respond to a challenge." The crucial political task of the future will be to bring together the people within nations so they can act as one in solving national problems and to bring together the nations of the world so they act as one in solving the world's problems.

A Contract of, for, and by "We the People"

When it comes to finding commonsense solutions to concrete problems, we the people are better, wiser, and more united than the politicians who represent us. The platforms of the political parties are drawn up to satisfy the entrenched self-interests of their most rabid members and therefore emphasize differences, rather than commonalities. In contrast, opinion polls consistently find broad majority consensus among the American people even on seemingly contentious issues that are stuck in the gridlock of Washington polarization. It is the great failure of our system of representative democracy that it is no longer accurately representative—the people are far less polarized than the politicians who purportedly represent them. And the difference has increased dramatically with time. Fifty years ago, polarization among the populace and polarization among the politicians were about equal. Polarization in the public has increased just a bit since then, but polarization among politicians has gone straight through the roof. We can, and must, force politicians to better reflect what the American people want and need, not what big money can buy.

We are still, and hopefully will remain, one people—once you eliminate all the divisive propaganda, ideology, and obfuscating rhetoric. The vast majority of us are practical, and want problems solved in a less ideological, more bipartisan way. Congress should be representing the general will of the people of the United States (as measured, albeit imperfectly, by opinion polls) rather than the narrow self-interests and radical ideologies of aggressive minorities (as measured by campaign contributions). Surely there are differences in how goals are to be achieved, but these are much smaller than meets the eye once we get past jargon to discuss concrete facts. The following "We the People Contract" reflects what nonpartisan pollsters find to be the American majority's view on crucial issues. The exercise reveals that the Trump platform serves only a minority interest and is far out of sync with what most of us want and need.

We, the people, want our government to:

- Accept the reality of climate change and reduce greenhouse gas emissions.[12]

- Emphasize alternative energy over oil and gas.[13]

- Raise taxes on wealthy individuals and corporations and eliminate tax loopholes to ensure that they pay their fair share.[14]

- Follow policies that promote a more equal distribution of wealth.[15]

- Protect Social Security.[16]

- Develop a simple, single-payer health insurance system modeled after Medicare.[17]

- Preserve Medicare. Government has the responsibility to ensure health coverage for all.[18]

- Negotiate lower prices for prescription drugs.[19]

- Reduce government waste and deficits.[20]

- Initiate campaign finance reform.[21]

- Be less partisan and gridlocked.[22]

- Improve the educational system.[23]

- Raise the minimum wage.[24]

- Lower tax rates for businesses and manufacturers that create jobs in the United States.

- Put people to work on urgent infrastructure repairs.[25]

- Enact a federal jobs creation law that would spend government money for a program designed to create more than one million new jobs.[26]

- Establish stricter policies to prevent people from overstaying their visas.

- Allow those born in the United States to illegal immigrants to remain here.

- Establish a way for most immigrants currently here illegally to stay legally.

- Keep unqualified illegal immigrants from receiving government benefits.

- Not build a wall between us and Mexico.

- Take in carefully vetted civilian refugees escaping from violence and war.

- Encourage more highly skilled people from around the world to immigrate to the United States to work.[27]

- Lower health-care costs.[28]

- Support continued federal funding for Planned Parenthood.[29]

- Keep abortion legal.[30]

- Reduce the prison population.[31]

- Decriminalize drug addiction and mental illness and provide adequate treatment for them.[32,33]

Having a wide consensus on broad goals doesn't mean that it will ever be easy to reach consensus on how to get there. But we must avoid "solution aversion"—denying problems exist because we disagree on how to solve them. Republicans deny the obvious evidence for man-made global warming (even though many know in their hearts that it's true) because they don't like solutions that threaten ideological values or financial self-interest. Studies show that they are more willing to acknowledge climate change as a threat when the solutions involve the "free market" (for example, technological advances) rather than government regulation or higher taxes. Getting beyond trigger words sometimes makes it easier to recognize reality—for instance, substituting "climate change" for "global warming" brings more Republicans around to accepting that the issue is valid. If we are to confront realities disguised by societal delusions, we must avoid jargon, buzzwords, stereotypes, innuendos, "dog whistles," and other distractions. The fact that compromises on solutions may be tough to forge doesn't justify running away from the problems that need fixing.[34]

One of the most polarizing attitudinal differences between

"conservatives" and "liberals" is how they understand the balance between individual effort and life circumstances in determining our fate. Many of my conservative friends share the conviction that they have earned their success and that people who are less successful just haven't worked hard enough. Most of my liberal friends and I are likely to feel we were more lucky than deserving, and would have done much worse had fate dealt us a poorer hand. We must bridge this gap. Success comes from working for it, but it also comes from the luck of birth and circumstance. Everyone deserves a fair deal but no one should expect that "society owes me a living."

Impeaching Trump Is Not the Answer

Trump is just a superficial symptom, not the underlying disease—just the clownish face of the sinister and well organized radical right. He would not have been elected president and would not be kept in office now were our country not so polarized by destructive color-war politics. Many Trump haters and Trump fearers entertain the seductive hope that Trump will self-destruct—get impeached by the House and then removed from office by the Senate. Or, perhaps, that he might quit the presidency in a huff, unable to tolerate the mounting criticism and constant investigative scrutiny occasioned by his continuing missteps. I yield to no one in my fear of Trump, but would much prefer he hang on (by the skin of his teeth) for the entirety of his four-year term, each day offering new lurid headlines demonstrating his personal idiocy and the unfairness and impracticality of his policies.

With Trump conveniently out of the way, Pence and his co-conspirators would function much more effectively to dismantle our governmental institutions; destroy the climate; rob the many to enrich the few; and assert white male, nativist hegemony. Trump got the Tea Party into the throne room, but Pence and Ryan would conduct themselves more plausibly while driving its same dreadful agenda. And they are also even less indebted to the legitimate demands of the populist constituency that propelled Trump into office (i.e., providing employment and healthcare to those left behind by our economy), while much more indebted to the big money boys and to the self-righteous, "religious" hypocrites. Undistracted by daily revelations of shady dealings with the Russians, by conflicts of interest, and by maladroit tweets, the more professionally functioning White House of President Pence, with the other two government branches in tow, can steamroll its ultra-right wish list of retrograde legislation (and perhaps to win re-election in 2020). It doesn't get a bit better should Pence be brought down by his close association with Trump's high crimes and misdemeanors. The presidential line of succession goes to the Speaker of the House, Ryan. And if Ryan also falls, we are stuck with President pro tempore of the Senate, the estimable Orrin Hatch. Then on to the Exxon nominee—Secretary of State, Rex Tillerson. A rogue's gallery of fools and knaves.

Our long, slow societal slide into delusional thinking was a long time in coming. Removing the frothy head won't eliminate the poison in the beer. Trumpfoolery has its uses. Better to keep bad boy Trump around, allowing him the irresistible opportunity each day to do or say something really stupid that will stall the noxious GOP agenda and drown the hopes of GOP candidates in swing districts. My paradoxical preference for Trump versus his potential inheritors assumes, of course, that he doesn't, during his remaining time in office, fecklessly press the nuclear button, or involve us more deeply in unwinnable wars, or try to establish a dictatorship. There is no denying that Trump is by far the biggest short-term risk in the history of our country; but Pence/Ryan can do far more irreversible damage in the long term.

Trump's troubles certainly do present the Republicans with an unpleasant conundrum. The interests of Koch types can be much better served by their own less independent and more reliable flunkies (Pence or Ryan)—but if they leave too many fingerprints on the process of dumping Trump, he will weave the GOP into a paranoid narrative that they are part of the "deep state" that engineered his downfall. Embracing the aggrieved martyr role, Trump could then whip his base into an enthusiastic third-party insurgency, gutting the Republican Party and scuppering its chances of holding onto Congress and the presidency. Republicans stand by Trump, despite his absurdity, out of fear not love. They will abandon him, as they did the much less dangerous Nixon, as soon as his poll numbers drop. Trump in post-impeachment retirement would do what he does and loves best—presiding at rowdy hate rallies all across the country and perhaps starting a hate media business. Trump might be even more permanently and dangerously disruptive of democratic comity in this role than as lame duck, clown commander-in-chief.

Winter Soldiers

These are the times that try men's souls. The summer soldier and the sunshine patriot will, in this crisis, shrink from the service of their country; but he that stands it now, deserves the love and thanks of man and woman.
—THOMAS PAINE

These words were written in the months after the birth of the United States. We were losing the early battles of our war for independence because of high rates of desertion—our citizen soldiers still feeling a much stronger call of duty to their families than to their newly founded, and still mostly unformed, nation. We persevered through the Revolutionary War because of the brilliant leadership of George Washington. We persevered through the even more bitter Civil War because of the greatness and goodness of Abraham Lincoln. We persevered through the gut punch of a severe economic and psychological depression because of the confidence and creativity of Franklin Roosevelt. All three men were acutely aware that the United States is an exotic plant, a tenuous and fragile experiment. None took the gift of democratic government the slightest bit for granted.

The United States has also survived many bad presidents, but never one so disrespectful of our fundamental institutions and values. We have often betrayed our ideals of freedom and equity, but never have we dismissed them so cavalierly. With Trump we have touched the bottom of every societal delusion; hopefully we will now bounce back with a better grasp on reality and renewed motivation to reshape it for a better future. I doubt we will return to the default pre-Trump status quo—more likely, we will either descend into a deeper dark age of tyranny and ignorance or we will emerge stronger, wiser, and more immune to the societal delusions that nurtured Trump and that he, in turn, is now nurturing.

I have always been politically passive, a skeptical and bemused neutral observer. Why even bother voting, when any given vote counts for naught? Why be loyal to either party when both are bought and paid for? Why engage in public citizenship that distracts from private responsibilities and private pleasures? I was among the substantial minority of our citizens who took citizenship for granted, who benefited from democracy without participating in it. Trump was elected president in 2016 because 40 percent of eligible voters did not vote—some too lazy, others indifferent to citizenship responsibilities, and too many the victim of cynical efforts to keep them away from the polls.

Now is again a time to try men's souls. Actually, way past time. We screwed up this past election. We elected a bad man, instead of a decent woman. The ship of state is foundering. The people are divided and confused. The stakes are high. The time is short. What Benjamin Franklin said is still true today: "We must, indeed, all hang together or, most assuredly, we shall all hang separately." The will of the people will be served only if the people express it loudly and express it clearly.

Our next three chapters describe the kind of world we would have should rational thinking replace societal delusion.

Chapter 7

Sustaining Our Brave New World

We exist in a bizarre combination of Stone Age emotions,
medieval beliefs, and God-like technology.
—E. O. WILSON

Sustainability depends on the interaction of just four variables: population, consumption, technology, and cooperation. Multiply world population by per capita consumption and you can estimate levels of resource depletion and waste generation. Technology determines how efficient we are at producing things and at getting rid of wastes. And cooperation will be necessary if we are to control population, equalize and tame consumption, and use technology to help rather than hurt sustainability.

At present, all four trend lines are going in just the wrong direction, converging on ever more unsustainability. In the past thirty years, world population has exploded by two billion,[1] while the number of people living in poverty has been reduced by one billion.[2] More and more people moving up the economic ladder means an enormous growth in overall consumption. So far, technology is doing more harm than good—by increasing the efficiency of extraction of irreplaceable commodities, it has promoted continued population growth, increased consumption, resource depletion, and waste accumulation. And the world seems to become less, rather than more, cooperative as there are more mouths to feed from a shrinking pie. There is no one grand solution to our sustainability challenge; there must be tens of millions of interacting small solutions, each in different ways reversing the trend lines in population, consumption, technology, and cooperation.

Population

There is great news on population control, pretty bad news, and really deadly news. The great news is that birthrates have already dropped dramatically in most parts of the world. The global average fertility of a woman was 5 kids in 1960 and is now down to only 2.5 kids, just a bit above the replacement level of 2.[3] Until recently, demographers optimistically predicted that reduced fertility would spread worldwide, with population peaking below 10 billion within the next three or four decades. Surely too many people, but perhaps manageable and thankfully finally under control. But the pretty bad news is that population growth is not slowing as predicted; the figures have been revised upward and don't have a clear peak. The world certainly can't afford anything above replacement fertility and would be a far better place if population were downsized in a gradual and benign way. Left to our own expansive devices, we risk a cataclysmic culling when our resources run out and our wastes accumulate.

And then there is the deadly news that the countries that can least afford population growth have birthrates that are crazily above replacement, while the wealthy countries have birthrates that are well below. All of the world's worst trouble spots are troubling precisely because of their unprecedented and unlivable crowding. In just seventy years, population in the Arab world has more than quadrupled, from 104 million to 450 million, and it is projected to reach a startling 700 million by 2050 (because the population of young women is so large and their birthrates continue to be so high). Most countries in sub-Saharan Africa still have birthrates of between five to seven children per woman, with Niger topping all at 7.6. Pick a country in the world that is suffering from a civil war, a massive migration, a famine, an epidemic, a massive die-off from natural disaster, or insufficient water and you will invariably find it to have astoundingly high fertility. Sample rates: 6.8 in Somalia; 5.4 in Sudan; 5.5 in Afghanistan; 4.6 in Yemen; 6.2 in Burundi; 6.3 in Congo; 6.5 in Mali; 4.2 in Gaza; 3.4 in Iraq. The routine sight of a mother in a refugee camp with six starving and thirsting kids offends the mind and breaks the heart. India is the most important population player in the world because it is second in number of people and until recently had a high birthrate. Fortunately, a combination of an active birth control program, an increasing middle class, and migration to cities has reduced rates from 3.6 in 1991 all the way down to 2.3 in 2013.[4] The trick will be to get other developing countries to follow this example as soon as possible. Unrestrained population growth guarantees subsequent Malthusian disaster.

Fertility rates in the developed world were once also high, but have now dropped precipitously, often considerably below replacement levels (especially in the Catholic countries of Europe, which apparently don't take the church's foolishly retrogressive reproductive policies very seriously). The drops, except for China, were a voluntary response to a combination of economic pressures, urbanization, cramped space, reduced infant mortality, children becoming more financial liability than asset, availability of contraception and abortion, changed cultural attitudes, and, most important, women being better educated, entering the workforce in numbers equal to men, marrying later, and having more control of their reproductive fate. China took a more active role in controlling population in the late 1970s with a much-maligned but essential one-child policy that prevented a population increase of several hundred million babies (who, of course, would have grown up to have their own babies). We may not like the authoritarian way China went about reducing its population explosion, but the harmful consequences of all those extra babies would have been far worse—not just for China but for the whole world.

Unfortunately, many low-fertility countries are now encouraging their people to have more babies, and birthrates are rising again in much of the developed world, including China.[5] Policies promoting population growth partly arise from nationalistic fears of losing political and economic influence if numbers drop—an irrelevancy if all countries readjusted population downward at equivalent rates. More pressing is fear of a demographic trap—that is, having an aging population, with too few younger working people left to support too many elderly retirees. Incentives to produce bigger families vary from country to country in ways you might expect: in Japan, it's money; in Denmark, it's sexy commercials (a gorgeous blonde saying "Do it in Denmark); in Finland, it's a goody bag; in Russia, it's a "day of conception" that enters you into a lottery for an SUV; in Singapore, it's "civic duty" in the bedroom. The fallacious logic is that a constant baby boom is needed to fill the demographic hole created by the aging population[6]—far better to correct the population imbalance in one generation than to provoke unsustainable population growth over many. Countries should wait out the hole, fill it with migrants or machines, or eliminate it by extending the retirement age of oldsters. The arbitrary retirement cutoff at around age sixty-five was established at a time when most people died just a few years later; it is set too low now that life expectancies have become fifteen years longer and oldsters are much healthier.

And we must change cultural values that regard childrearing as an essential human experience and a requirement for happiness. Among all groups surveyed, single parents were most likely to say they were "not too happy" with their life.[7] Straitened financial and environmental circumstances often make a kid less a bundle of joy, and more a pain in the ass. Surveys show that mothers much prefer talking on the phone or watching TV to playing with their kids, and that parental happiness decreases the more time parents must spend with their children. The overall concept of parenting may be great, but the day-to-day doing of it can be trying, especially if childrearing is sandwiched into a schedule that is already too busy. And as the saying goes, a mother can only be as happy as her unhappiest child. Finally, the last thing an already overcrowded world needs is ever more sophisticated fertility clinics that create multiple births and encourage women to have children into their late forties and even their fifties.

The wide variability in long-term UN world population predictions reflects all the Malthusian uncertainties of the far-off future. Estimates for 2150 range from just three billion all the way up to twenty-five billion.[8] The low end assumes a catastrophic die-off in a chaotic, apocalyptic world. The high-side prediction assumes catastrophic crowding that could not possibly be stable. My own guess is that over the very long haul, the world can sustain only something close to the lower number. The question is whether we get there by rational and gradual planning, or by killing one another off. The developed world has convincingly proven that proper population control is possible; now the question is whether we have the good sense to make it probable through the developing world.

What can be done? The first step in solving a problem is facing it. The news is filled daily with the latest disaster occurring in Afghanistan, or Iraq, or Syria, or Yemen, or Egypt, or the Congo, or Sudan, or Somalia, or any of a dozen other countries maintaining impossibly high fertility rates in the face of poverty, unemployment, failed crops, and declining water supply. Almost never is there any discussion of the underlying population demographics that are most certainly driving the disaster. Reporting on terrorism rarely mentions that it follows predictably from a world overpopulated with young men, crowded together with no prospects for job or marriage, and having nothing important to lose except their lives. Or that the bulge of young women having multiple babies is a population time bomb. We must break the taboo on openly discussing that excess population is the underlying disaster that causes many of the other disasters. It is never a simple Shia-versus-Sunni conflict—it is almost always too many Shias fighting for dwindling resources with too many Sunni. And so it is with Tutsi versus Hutu, Pashtun versus Tajik, Tamil versus Sinhalese,

Jew versus Palestinian, and so on. Nearly every war and civil war is a battle between overpopulated, patriarchal tribes (continuing to multiply fiercely) fighting neighboring tribes (also multiplying fiercely) for rights to rapidly diminishing land, food, water, and other resources. The wars, migrations, and famines will continue to be the only control over otherwise uncontrolled population until and unless people come to their Malthusian senses.

How much can we tame the mad fertility so cherished by our cultural traditions (slavishly serving, as they do, the selfish interests of our DNA, not our own sustainable survival)? That it won't be an easy task is clearly illustrated by two stories pulled from today's newspaper (almost random stuff; similar headlines seemingly appear every day). The president of Turkey says it is unpatriotic and unfeminine for women to work; she must stay home and have more children or she is "a half person." This in a country already massively overpopulated with native Turks and Kurds and overrun by three million Syrian immigrants.[9] Go figure. And the World Health Organization (WHO) is equivocating as to whether potential parents living in Zika zones should delay pregnancy—first suggesting yes, then taking the politically correct and culturally sensitive route of not being so sure.[10]

We can't beat overpopulation if we remain afraid to talk openly and frankly about it. And discuss it we must if we are to minimize constantly recurring Malthusian disasters. Warring tribes around the world may think they are fighting an existential struggle with one another, but they are actually waging an unwinnable battle against their own overpopulation and the maladaptive cultural traditions that promote it. Perhaps it feels (and is) culturally insensitive to make this plain, but it is also essential. The irrational religious, ethnic, and tribal passions stirred by any hint we should control population must be confronted if they are to be made less passionate and less irrational. The world needs persistent and pervasive lessons in demography—at diplomatic meetings, in news reports, in schools, and hopefully even in churches, temples, and mosques. Having more kids means condemning the kids you have to a much worse life. Quality of life beats quantity of life. A small family that succeeds is better than a large family that starves. The sanctity of life means holding precious the kids we have, not forcing them into grim competition with kids we cannot adequately provide for.

Reproductive fundamentalists of all religions need to wake up to the suffering inflicted by overpopulating our already cramped and conflicted world. The United States must reverse its tight restrictions against funding family planning as part of foreign aid. And the Catholic Church must reconsider its misguided crusade against contraception and abortion. Supposedly championing the dignity of life, it is instead guaranteeing the most undignified and miserable of lives for starving, sickly, war-torn, and migrating hordes. Politically influential evangelicals should not be empowered to block our government from funding family planning in the many countries around the world that desperately need it. Population control must be the number one priority in the world and our highest moral imperative. Morality is always the first casualty of a world that cannot provide for all its people.

The Gates Foundation provides a good model of flexibility. Its original hope was to facilitate population control only indirectly by improving health, reducing mortality, and educating women. When these measures failed to stem the out-of-control population growth of Africa and South Asia (and may instead have contributed to it), the foundation began and has since expanded programs to educate about family planning, and provide the means for accomplishing it, to hundreds of millions of women.[11] Governments and foundations around the world must follow suit. If we don't adhere closely to Malthus's advice by instituting benign birth control, surely we will be visited with Malthusian disasters of increasingly biblical proportions.

World population will eventually be downsized one way or another as we run out of resources. We can stick with our delusional population optimism and eventually suffer all the dire consequences of war, pestilence, drought, and famine, or we can adjust gradually and gracefully by beginning a concerted, well-

financed, worldwide effort to promote smart reproductive pol-
icies. Many people retain a firm religious conviction that birth
control is a sin. It seems to me that delusionally denying the need
for birth control is a far greater and more unforgivable sin. Kids
should not suffer being born into a world that simply cannot pro-
vide for them.

Trump couldn't care less about contraception and abortion,
one way or the other. But he made a devilishly clever deal with
the opportunistic evangelical leaders who form an important
wing of the Republican Party. They would overlook the fact that
he is among the least moral of men, and he would deliver them
strictures against birth control (restricting abortion, defunding
Planned Parenthood), not only in the United States, but also in
the strings we attach to aid programs across the world. To seal
his promise, he was forced to accept Mike Pence as vice presi-
dent—a politician long in the employ of the most radically ex-
tremist fringe of the evangelical movement. Trump is famous for
breaking promises, but Pence has kept him honest on this one. A
small cabal of old white men have exerted their extremist control
over the reproductive rights of millions of young women—and
in the process, have put our world on the course to a devastating
Malthusian crash. In their inexplicably tepid support for Hillary
Clinton, women allowed Trump to steal the presidency. They
have begun since to organize to express their displeasure at his
cynical enforcement of draconian reproductive policies. But it
may be too little, too late. It is no exaggeration to say that the
fate of the world may be determined by how the 2018 and 2020
elections affect birth control policy and population growth.

Consumption

Bacteria are cultured on petri dishes—small containers holding
an abundant, but finite, food supply. Settle just a few bacteria

on an immense quantity of bacterial yummies and the results are absolutely predictable—wild multiplication and unrestrained consumption until the dish so fills with bacteria that it becomes empty of food—and the colony dies off completely. "Catabolic collapse" is a term used to describe the human version of bacteria consuming themselves out of existence on a petri dish. Historically, all complex and successful societies come eventually to live beyond their means, making and consuming more than they need or can possibly sustain, and then they collapse as resources are exhausted. It is a constant finding in archaeology that civilizations reach their apex of productivity just before they disappear—the thickest level of junk in any midden is usually the last.[12]

No society ever produced or consumed or discarded more junk than ours. When I was a kid, a toy was a rare and precious thing, a treasured companion through childhood. In contrast, this past Christmas, at my son's house, resembled a lavish potlatch, with mountains of stuff that will reside only briefly in the house before being relegated to the garage, on its way to the garbage bin. Kids don't need or benefit from all the excess kid toys they get. And adults don't need or benefit from all the excess adult toys they have been taught to crave. We will last longer as a society if we can cut down our rate of consumption.

Unfortunately, our economic policies are all based on the fatally flawed opposite assumption—that constant growth is not only an inherent good, but also a fundamental essential for national survival. Any interruption in growth is reviled as "recession" or "depression" and reversed by desperate fiscal and monetary measures to get people to spend and consume more. Too much (70 percent) of our GDP is derived from consumer spending of often useless products; too little for infrastructure projects and research that would lead to a more efficient and sustainable world.[13] We build too many cars, too few systems of public transport; do too much market-driven, meaningless drug research, and too little research on clean fusion energy to replace disappearing, dirty fossil fuels; and buy way too many silly toys from China. The entire advertising industry is devoted to seducing the orgy of mindless consumption brutally satirized in Huxley's *Brave New World*. Planned obsolescence is the name of our economic game. All great for corporate profit, but unsustainable and essentially unnecessary if our goal is simply to make people happy.

As the developing world raises its people out of poverty and into modernity, it must increase its consumption. This is fair and right, but sustainable only if there is compensatory decreased consumption in the developed world. We must switch our attitudes, institutions, and economy away from growth and consumerism and toward sustainability and contentment—stop producing vast volumes of things we don't really need, and instead focus on low-cost/high-quality happiness from the things we can afford. A simpler life is very often a happier life.

Retooling Technology

The tremendous efficiencies afforded by modern technology present us with three simple choices: (1) make more and more stuff; (2) work fewer and fewer hours; (3) have more and more people unemployed. John Maynard Keynes, the most brilliant economist of his time and a savior of capitalism during its darkest moment, enthusiastically predicted in 1930 that mankind would soon require only a fifteen-hour workweek, providing added leisure to cultivate our interests. This was a reasonable guess given that rapid technological advances and rapid reduction in the workweek had both been achieved over the previous fifty years.[14] Herbert Marcuse, a fierce critic of capitalism, saw an ever-shortening workweek in moral terms—as the only way to protect mankind from shallow, consumerist "one-dimensionality." "The people recognize themselves in their commodities; they find their soul in their automobile, hi-fi set, split-level home, kitchen equipment." There is no time permitted for culture, no room for independence. If excess labor is flattening human life, less labor is essential to rounding us back into three dimensions. "The reduction of the working day is the first prerequisite of freedom."[15]

Keynes and Marcuse were both as smart as one can be, but both proved to be very poor predictors of the impact of technology on leisure—at least so far. Advancing technology has either enslaved workers even more or replaced them altogether. Each of man's technological revolutions has historically resulted in people working harder than they ever did before, because the increased productivity always gets turned into more products and more people, not into more leisure. Hunters and gatherers before the agricultural revolution had more leisure than did the farmers who came after; the farmers before the Industrial Revolution had more leisure than did factory workers; and factory workers before the Information Revolution had more leisure than computer jockeys. Google has probably worked hardest to create a high-tech utopia that will recruit and retain the best and happiest people. Much-envied perks include: a beautiful campuslike work environment; free gourmet breakfast, lunch, and dinner; coffee and juice bars; free fitness center; free transportation to and from work; generous maternity and paternity leave—and don't forget the financial rewards, the tech toys, and the chance to work with other smart people. But there is no possible way to get away from the office when you are chained to it by smartphone and computer, are expected to be available 24/7/365, and know that your competitors are willing to work till they drop.

The crucial achievement of a sustainable economy must be everyone working less, but everyone working some. We need to employ our technology as Keynes and Marcuse envisioned—to make ourselves more interesting and interested people, maybe poorer in stuff but richer in time, wisdom, happiness, and relationships. Massive increases in productivity have, for the most part, been wasted—producing mostly useless stuff and/or in making workers useless. The biggest risk of computerization and taming GDP growth is massive unemployment, which has a devastating impact on life satisfaction and is an important risk factor for suicide. Spreading employment with a short workweek and taking better care of those who are unemployed are the best forms of suicide prevention and the most powerful contributors to national happiness. Interestingly, unemployment in the Nordic countries causes much less unhappiness, because they provide such a strong safety net.[16]

One great thing about capitalism is that the imbalances of the market can be fairly easily corrected with taxes, subsidies, and regulation. One terrible aspect of capitalism is the easy corruption of the tax, subsidy, and regulatory provisions to further advantage the advantaged and further disadvantage the disadvantaged. Our tax and regulatory codes have mostly been written by industry lobbyists, not surprisingly to favor short-term industry interests at the expense of the public good and of future generations. The good news is that simple adjustments could realign industry incentives in the right direction. Companies should be taxed not just based on current income, but also on the direct opportunity costs their activities lay on the future (think carbon tax). Generous subsidies to the fossil fuel and agribusiness industries should be transferred to competitors who can produce clean, sustainable energy and food. Tax breaks should go to those technologies that create the greatest efficiencies in reducing waste, enhancing infrastructure, and producing the most lasting social benefit. We also need to end planned obsolescence by rewarding technologies that make quality products that last, not disposable junk that must constantly be replaced. We need to become an economy much less dependent for its health on short-term consumer purchasing and much more dependent on long-term industry investment, now lagging badly because corporations are hoarding trillions of dollars (mostly offshore to avoid taxes).

New technologies extract fuel, metals, and food from the earth with unprecedented efficiency—but also with the unintended consequences of more population growth, fewer remaining resources for future generations, ever more pollution, environmental degradation, and fewer jobs. We must redirect technology so that it is cleaning up the world, not despoiling it. Tax and policy reforms may seem dreamily unrealistic, especially given the Trumpian corporate and big-money capture of Washington's taxing and regulatory powers. Dire exigency will before long conquer inertia and sensible leadership will someday initiate more equitable and sustainable policies. The troubling question is whether this will happen in time.

Cooperation

Ants within a colony have an inborn, immediate, urgent, compelling, and complete need to cooperate with one another. Because each ant is so closely related genetically to every other ant in the colony, ant DNA programs complex behavioral algorithms that ensure an "all for one, one for all" approach informing every action of each ant in the entire colony, from its birth to its death. There are no civil wars within an ant colony, no coups, no partisan wranglings, no selfish haggling, no controversies, no contract disputes, no reneging on agreements. An ant colony works as a "superorganism"—no isolated ant can survive alone, cooperation is all, and competition within the colony is unthinkable.[17] Different ant colonies fight one another fiercely to the death, but each colony acts like one organism. There is no more individualism within an ant colony than there is within the organs of the human body. When our bodies function healthily, it is because every cell knows its place and cooperates fully with every other cell. We call it cancer when the cells of any of our organs take it upon themselves to selfishly multiply at the expense of other cells. Lack of cooperation in a society can degenerate to a kind of social cancer.

We expect complete cooperation within our internal organs, but don't expect complete cooperation among different humans, even those who are closely related. For better, and now very often for worse, natural selection rewarded design specifications in humans that are very different from those in ants—devotion to the common good is not nearly so much a part of human nature as it is of ant nature. We are striving, competitive creatures with only limited capacity for cooperation and sharing, even within and certainly outside the immediate family circle. Ants evolved to be so inherently social, and so unselfishly noble, because they are such close kin with everyone else in the colony. Humans share many fewer genes with kin—we have some abilities toward altruism, but are inherently much more selfish and individualistic, more every-man-for-himself, programmed to fear that nice guys will wind up finishing last. Our kinship altruism exists, but it is much more limited in its strength and circumscribed in its horizon. Our noblest intentions are often defeated by selfish concerns, contentious bickering, and divergent views of what is fair and right and good. That's why human experiments in antlike communal living (the kibbutz, utopian villages) eventually flounder and fail.

Naturally enough, we have an ambivalent attitude toward ants—admiring their ingenuity in working together, but also feeling dread at the thought of ourselves being so automated for the common good, so lost in the crowd. Our genes allow us—or rather, direct us—to have at least some of that Frank Sinatra "I gotta be me" feeling. But, for historical and geographic reasons, its strength varies greatly from culture to culture. Societies like China and Japan that are genetically homogeneous and early experienced the pressures of crowded living developed a Confucian, consensus-building approach that greatly favored cooperation over individualism. Geographically and genetically more disparate Europe rewarded more individualistic tendencies and these found their fullest expression in the genetically heterogeneous, and once wide-open, United States.

Now that the world has shrunk and become as crowded and interdependent as an ant colony, our needs have also changed. Unless we can overcome the worst consequences of our rugged individualism, our days as a successful species are numbered. We can't expect humans to ever cooperate with antlike uniformity, but we are doomed if we don't learn something from antlike co-operation and apply it to our current situation.

Confucian wisdom must play a much larger role in all of humanity's endeavors. Never in world history has worldwide cooperation been more imperative or easier to obtain. Imperative because all our existential threats arise from global threats and all will most certainly require global cures. We are all in the same small and very crowded boat and can no longer feel safe protecting our family and local tribe, letting the rest of the world be damned. We sink or we swim together. Fortunately, all the global institutions required to solve our problems are up and running (for example, United Nations, International Monetary Fund, World Health Organization, World Bank, as well as thousands of nongovernmental organizations [NGOs] and philanthropic foundations). The corporate world is no longer divided by national boundaries and business works best when cooperation is greatest across them. Europe has transformed itself from the world's bloodiest battleground to its largest-ever united market.

Though the need for worldwide human cooperation has never been greater or more apparent, its fulfillment so far has been disappointing. The United Nations is anything but united; the European Union has been anything but a union in response to its migration, terrorism, or currency crises; the WHO has been poorly organized in fighting epidemics; the IMF and World Bank are widely reviled; and tensions have worsened with Russia and China.

Stress can strengthen a system or fragment it. So far, we have seen mostly fragmentation, with Trump leading the demolition team. But it is still early days and I expect this to change. Soon it will be impossible to ignore that world population growth is no longer supportable, that oil and gas are running out, that we have changed the climate, and that the air is hard to breathe. The question is whether the wising up will come soon enough, and whether the cooperation will be great enough, to get things into sustainable balance before they are too unbalanced to be set right.

Sustaining Nature

When I was a very young child, my family moved to a leafy suburb of New York City that was then about equally divided between greenery and concrete. My walk to elementary school meandered through a small, but to me enchanting, field of trees, wildflowers, butterflies, fireflies, and imaginary cowboys and Indians. Then the developers moved in and began eating up the open spaces. By high school, looming apartment buildings had created concrete canyons dotted with postage-stamp-size lawns that provided the only pathetic reminders of a lost natural world. I felt claustrophobically smothered except when we could escape briefly to the open air of beach, park, or mountain. My happiest moments were then, and still are, natural moments.

Woody Allen urban types—the kind who crave concrete and detest dirt, wind, sand, and sun—certainly exist, but they are a minority. Thousands of generations of our ancestors lived in nature and made their living off it. To find food and avoid becoming something else's meal, they had to be terrific naturalists with intimate knowledge of botany, zoology, geography, and climate. And to be good at your work, you have to love it. Our species is pre-adapted to find nature beautiful, to feel attracted to living things (what E. O. Wilson calls innate "biophilia"). Potential ancestors who didn't enjoy walking in the woods, or swimming in the lake, or a nice sunset, or tracking prey would lack the zest needed for survival. Their genes could not compete with those of nature lovers. Farm life is only 10,000 years old, city life 6,000, factory life 200, computer screen life 30. We can adapt to any environment if we must, but unnatural environments are not in our genes.[18] As John Muir put it, "The clearest way into the Universe is through a forest wilderness."[19]

There is a growing disconnection between our enduring need for nature and our rapidly disappearing access to it. We love zoos, botanical gardens, national parks, camping, fishing, hunting, gardening (and my personal favorite, beach bumming) because they put us back, at least for a while, where we belong. In much of the developing world there has been a recent massive migration from rural living to densely packed urban quagmires. In much of the developed world, there has been a wrenching shift of attention away from real reality and on to the virtual reality of a computer screen. Kids have never spent less time playing in nature and most people in developed nations are trapped indoors for most of their lives. We will not feel the appropriate responsibility to protect nature and nature's creatures if we no longer know and revel in them. We need to get back to nature, to live and breathe its reality—and get our kids' faces away from computer and touch screens.

Earth, once our mother, has now become our child, dependent on us for protection and stewardship. It took our species hundreds of thousands of years to master Mother Nature, but only a few hundred years to despoil her. Anyone who sees the forest for the trees can tell the forest is burning—quite literally here in Southern California, where I live; figuratively for the entire world. The biggest threats to the natural world are the fatal combination of overpopulation, consumption, and economic expediency. We are familiar with the hypocritical refrains: "we have to clear this rain forest to feed the people and generate jobs" or "this pipeline will be essential for our economy" or "these environmental regulations will put us out of business." The economic calculations used to justify harming the environment are always based on short-term profitability that benefits only a very few people for only a very few years, ignoring the long-term costs paid by all the rest of us over the many centuries. Big-buck commercial interests spend tens of billions of dollars each year buying politicians, discrediting science, and scaring the public with threats that jobs will be lost and the economy will fail should we follow responsible environmental policies. Corporations and the very rich have turned conservation into an ugly partisan political issue, even though it should be seen as an obvious benefit to all mankind. Secular big business has also formed a strange but powerful alliance with those in the radical religious right, which has become preoccupied with micromanaging and policing morality, and is mostly indifferent to the Bible's call to be good stewards of the earth (seen by them as God's job). Fortunately, the more enlightened religious groups have gone "green" in recent years—recognizing their responsibility to protect as God's gift our beautiful natural world.

Although alive and well, the environmental movement is fighting an uphill battle against big money and vanishing time. E. O. Wilson, one of its most learned and eloquent spokespersons, despairs, "We live in a delusional state. America, in particular, imposes a horrendous burden on the world. We have this wonderful standard of living, but it comes at enormous cost. To bring the world's seven billion people using today's technology up to the level of the average American will require four more Planet Earths." We don't have four more earths and never will. Survival requires our being a lot smarter and kinder in our use of our one and only, lonely planet. Wilson's solutions will be familiar to you: setting aside large nature conservatories in biodiversity hot spots all around the world; controlling population by educating and empowering women; radically reducing energy consumption and radically increasing use of environment-friendly sustainable resources; and a new green revolution to produce more calories on less land.

Earth, in the geologic time scale, is impervious to our petty invasions. Take man out of Manhattan and it will restore itself to lovely forest in just a few centuries.[20] Cambodia was one of the richest and most populous places in the world in the twelfth century, but you wouldn't guess it now because vast cities are all buried within a regrown jungle. But in the short human time scale, we can hurt nature badly enough to severely hurt ourselves. Nature is the canary in the cave—destroy it and we are next. Sustaining it is a great long-term economic investment, a moral imperative, and a labor of love.

Dusk or Dawn?

Our species is at a tipping point and in a twilight zone—we are either at dusk, just before entering a new dark age, or at dawn, just before leaving a past one. Without question we can and must reach sustainability, but it remains frighteningly uncertain whether we will. Current policies could not seem less promising—materially unsustainable, morally bankrupt, incredibly dumb. Many countries have put the tiller in the hands of reckless and feckless skippers, who are sailing our little ship into a perfect storm of overpopulation, greedy consumption, and fierce competition. Righting the course at the eleventh hour will be difficult and may seem a long shot—but must be done and almost certainly is not impossible. As Samuel Johnson put it: "When a man knows he is to be hanged . . . it concentrates his mind wonderfully." Our dire necessity must be the best mother of our reinvention. Once we achieve it, sustainability will be easy enough to maintain. The tough part is going from here to there—getting past our societal delusions and leaving behind the dystopic Trumpian dark age.

Trump is an irremediable creature of Stone Age emotions and medieval beliefs. His policies are a five-way disaster, impacting negatively on each dimension required for a sustainable world. Trump is doing his best to increase world population by cutting funds for family planning, at home and abroad. He encourages consumption and discourages conservation. Trump fights the science and technology that might reduce waste and pollution and promotes projects that increase them. He is signing executive orders to sell off public lands and wildernesses to the highest bidder. And this president does not ever work and play well with others. Our only hope is that Trump's being the poster child for such colossally stupid policies will help discredit them forever—not just in presidential elections, but, and just as important, in Congress.

Chapter 8

The Pursuit of Happiness

How much you don't need to be happy.
—MARLOW

Sometimes a brief encounter in a strange place sticks with you and changes the way you think about and live the rest of your life. About twenty years ago, I was vacationing on a tributary of the Amazon, four hours downriver by small boat from the nearest dingy oil town. I was holed up in a tiny eco-lodge that seemed likely at any moment to be drowned by the pounding rain or swallowed up by the encroaching jungle. The lodge had only one employee—a precise, efficient, little man from Vienna with a dapper mustache and charming accent. He served as manager, boatman, guide, chef, waiter, bartender, handyman, and cleanup crew—filling all these roles cheerfully and with brilliant aplomb. But he also stuck out as incongruously urbane—why was an educated guy like this doing mostly manual labor in the middle of nowhere?

People seem to feel compelled to pour out their life stories whenever I am forced to admit I'm a psychiatrist—even though listening to them is not really my favorite vacation pastime. But this man had a history that needed telling and was fascinating to hear. Spooling out his tale in the damp and lonely darkness, he sounded like the narrator in Conrad novels. Let's call him Marlow. Until two years before, Marlow had led the comfortably conventional life of a typical, well-to-do European intellectual. He was a civil engineer, married to a doctor, and had a teenage daughter. They lived in a large and lovely apartment filled with books, antique furniture, and a baby grand piano. He was a passionate collector of ancient coins and modern art, loved Mozart, and was fond of gourmet cooking and fine wines.

Then, without warning, Marlow's well-ordered life suddenly came undone. Encouraged by a foolish psychotherapist promoting reckless truthfulness, his wife made the guilty confession of having had occasional brief and casual sexual liaisons with his best friend. Although their last physical encounter was several years before, she was still filled with guilt and wanted Marlow's understanding and forgiveness. These were both beyond his purchase to give. Marlow's "overreaction" (as he now put it) included a frigid and bitter rejection of his wife, intense jealousy, murderous rage, deep sadness, and crippling self-doubt. He had unsuccessfully tried to drown his misery with a whole variety of prescription pills and street drugs, but more reliably found temporary relief at the bottom of a bottle of alcohol.

Marlow was not by nature a violent man and was not one given to staging embarrassing scenes. But he couldn't control his thoughts. Should he kill her, or kill himself, or kill his best friend, kill all three? He was haunted by vivid imaginings of his betrayers making passionate love and could not stand the sight, sound, smell, thought, or presence of either of them. He endured six excruciating months of distant, silent, rageful mistrust before deciding he must either leave his wife or kill himself. A plan emerged that accomplished both goals in a manner closely calculated to best protect his daughter from the inevitable pain and shame of losing him to suicide. Marlow announced he would take a two-week ecological and cultural vacation in Ecuador—to observe flora and fauna in an unspoiled habitat and to study native customs in remote villages.

This plausible cover story aroused no suspicion since he had previously made similar solo jaunts for the same purposes in Laos,

Madagascar, and Thailand. Marlow quietly but meticulously put his work, possessions, and personal affairs in good order, prepared his last will and testament, and set off on his suicide mission. In Quito, Ecuador, he engaged a bush-whacking charter pilot to fly him to the most remote jungle landing strip and then a boatman to take him as far upriver as he was willing to go—a full day's ride into what would become Marlow's own personal heart of darkness. A sham agreement to return for pickup four days later was the plausible disguise of Marlow's exotic suicide. He would instead walk deep into the jungle until he was irretrievably lost. The fates would decide the particulars of the inevitable endgame.

It didn't work out as planned. Indeterminate days later, a delirious Marlow was discovered by a seminomadic tribe of hospitable natives who maintained only minimal contact with the outside world. They gradually nursed him back to health and during the next year provided sustenance both to his body and to his soul. Marlow learned bits and pieces of their language, but it was their unconditional kindness and hearty acceptance of their paltry lot that cured his alienation and removed the appeal of suicide.

Marlow was now six months into his "halfway house" experiment of managing the little lodge. He was still not sure what to do with the rest of his life, but he was passionately sure he wanted to live it. Should he forever remain at this boundary place between his two previous lives or should he attempt a return either to his family or to his new friends in the tribe? This seemingly crucial life decision occasioned him little concern—Marlow now knew that peace of mind was centered within himself and was not subject to unpredictable external circumstance. Parting from him for the last time, I gave Marlow a hug and asked what he had learned from his extraordinary odyssey. Without missing a beat, he said: "How much you don't need to be happy."

This is, I think, the best advice to help our world stop living beyond its means. We must embrace choices that bring us sustainable happiness in the things that really matter and abandon the fake happiness that comes from pursuing excess consumption.

Marlow seemed the wisest and happiest of men. Trump must be among the unwisest and the unhappiest. The sacrifices we must make to become sustainable are not really sacrifices.

Pleasure Principle vs. Reality Principle

Maximizing pleasure and minimizing pain are the most basic and oldest of all behavioral motivators. The first living cells billions of years ago had the discernment to approach things that felt good, to avoid things that felt bad. It is the clearest possible confirmation of the continuity and conservatism of evolution that worms and flies, and the other humble creatures of the earth that first evolved nervous systems hundreds of millions of years ago, still use the very same neurotransmitter (dopamine) that we do today. And for the very same reason: to stimulate the search for the good things in life like food, drink, sex, and hanging out with other worms and flies. The amygdalae are headquarters for our reward pathway, but pleasure is important enough to rate rich connections everywhere, especially to the memory and decision-making centers.

Making good decisions requires managing the seductions of pleasure, tolerating the discomforts of pain, and having a realistic view of how much of each we can expect. Almost all philosophers and psychologists have had to address pleasure, pain, and their relation to our daily reality.

The Epicureans of ancient Greece and Rome were remarkable scientists, able to work out some of the most counterintuitive fundamentals of reality, two thousand years before Newton and Einstein. They understood that simple and indivisible atoms must be the basic building blocks of the universe; that these atoms move randomly and ceaselessly within a void, occasionally crashing into one another to form the complex stuff of the natural world; that atoms have position, number, weight, and shape; and that we perceive things via atoms traveling to our sense organs at great speed. Epicurean philosophy was strictly materialistic—no gods, no superstition, no utopian delusions. We have only one life and only one goal—to make the most of it by promoting pleasure and minimizing pain. And we can make the most of our lives only if we face reality straight on and without distracting illusions.[1]

Epicurus enjoyed his nice garden, Mediterranean breezes, good food and drink—but his definition of pleasure was definitely not Hollywood or *Brave New World*. Pleasure comes from the tranquillity of a modest life, limited desires, gaining knowledge of oneself and the world, good fellowship, and doing one's job. Death is not to be feared, because it is a state without consciousness, no different from our status (or lack thereof) before birth. Epicurus's last words show his calm and realistic appraisal of life and death: "I have written this letter to you on a happy day to me, which is also the last day of my life. For I have been attacked by a painful inability to urinate, and also dysentery, so violent that nothing can be added to the violence of my sufferings. But the cheerfulness of my mind, which comes from the recollection of all my philosophical contemplation, counterbalances all these afflictions."[2] A great model of equanimity and generosity of spirit for all of us.

Developed at the same time in the third century BCE, Stoicism and Epicureanism were major competitors—one focusing more on the management of pain, the other on the production of pleasure. But this was little more than the narcissism of small differences—both philosophies had a similar materialistic view of the world and how best to behave in it. The stricter of the two, Stoicism taught the value of controlling emotional reactions to the slings and arrows shot at us by outrageous fate. "Live according to nature"—if nature is guided by reason, why shouldn't our human nature strive also to be completely reasonable; equally indifferent to pain, sickness, poverty, passion, and good fortune? Stoics would have no patience with our societal delusions. Their advice would be to use reason to solve our planet's problems and not to flinch from the pain we may face in the process.[3]

The next great contribution to a materialist morality came two centuries ago from Jeremy Bentham. Influenced by the revival of ancient learning that informed the Enlightenment, he created a utilitarian calculus meant as a practical guide to individual moral judgments and societal decisions. "Nature has placed mankind under the governance of two sovereign masters, pain and pleasure. It is for them alone to point out what we ought to do, as well as to determine what we shall do. On the one hand the standard of right and wrong, on the other the chain of causes and effects, are fastened to their throne. They govern us in all we do, in all we say, in all we think." Pleasures and pains are to be measured as precisely as possible according to their strength, duration, predictability, immediacy, risks, and generalizability to others. They can be summed up for the individual and aggregated for the society. Public policies are not to be judged on abstract principles, but rather on their practical consequences—do they bring the greatest good to the greatest number, now and into the future?[4]

Utilitarianism is imperfect, but essential. Imperfect because it is impossible to measure utilities independent of value judgments— Hitler could make claim to being a Benthamite furthering the greatest good for Germany, while committing the worst atrocities against mankind. Essential, because there is no better available guide to individual behavior and public policy. The yet unanswered, but crucial question, is whether our fractious species can ever come together in deciding what are its most essential survival values, how best to measure their accomplishment, and which policies are most likely to increase world pleasure and reduce world pain, taking into account responsibility for protecting the pleasures and minimizing the pains of our species over the hopefully long arc of our future.

Sigmund Freud is now paying the price for being overestimated during his own lifetime, by being greatly *under*estimated in ours. Having a firm command of neuropathology and evolutionary theory, Freud intuited that human psychology would reflect the hierarchical layering of human brain structures upon those of our animal ancestors. Our unconscious brain functioning, mostly devoted to satisfying primitive instincts, follows the "pleasure principle" of immediate gratification. Our conscious brain function, in contrast, follows the "reality principle"—the ability to delay gratification, apply rational reasoning, and respond appropriately to the demands and opportunities of the external world. "An ego thus educated has become 'reasonable'; it no longer lets itself be governed by the pleasure principle, but obeys the *reality principle,* which also at bottom seeks to obtain pleasure, but pleasure which is assured through taking account of reality, even though it is pleasure postponed and diminished." Infants are pure pleasure system; psychological maturity comes with increasing ability to contain the pleasure principle with healthy doses of reality testing.[5] Freud's distinction anticipated Daniel Kahneman's later division of thinking into System 1 and System 2.

Societal delusions are pure pleasure principle—out of touch with threatening reality and unable to respond appropriately to it. Freud stated as his goal for individual therapy: "Where Id was, there Ego shall be." Similarly, the goal for our society must be to apply rational long-term planning to deal with real-world problems, not submit to denial and wishful thinking to facilitate short-term indulgences.

Trump is pleasure principle embodied—reality and truthfulness be damned. He has the attention span of a gnat, the temper tantrums of a two-year-old, and the impulsivity and grandiosity of a teenager. For most of us, reality eventually bites—experience teaches us that we are not the center of the universe, punishes us when we lack self-control, and forces us to grow up and follow the rules. Born with a golden spoon in his mouth, Trump is the classic spoiled brat who never learned how to subordinate his needs to the needs of others. Never burdened by the reality principle in his personal life, Trump is uniquely unable to apply the reality principle to presidential decisions. He governs from the gut, not the brain; makes up alternative facts rather than facing real facts; and refracts every event through the lens of his self-importance. We need to grow up as a society and apply the reality principle to the challenges of our future. President "pleasure principle" Trump is the wrong man in the wrong job at the wrong time.

Happiness Set Point

Transforming from a growth-dominated economy to a sustainable one will certainly create major disruptive consequences, both intended and unintended. Constant growth first became mankind's guiding goal with the advent of the agricultural revolution ten thousand years ago. Accelerating growth became the aim of the capitalist revolution four hundred years ago. Exponential growth began with the Industrial Revolution two hundred years ago and has been most fully realized in Silicon Valley today. The multinational corporations that now control the world live and die by quarterly earnings, conservation of the long-term future be damned. It won't be easy to learn to live within our collective means after ten millennia of reckless profligacy. Making the transition to sustainability will create unavoidable hardships and inequities that would be impossible to justify were the status quo not inevitably an even more perilous choice.

But there is one source of comfort among all the many uncertainties. Because more stuff hasn't made us more happy, less stuff won't necessarily make us less happy. Once income rises above a certain level, raising it further has little impact on happiness. The best explanation for this unexpected and counterintuitive phenomenon is the "happiness set point."[6]

Homeostasis is one of the most valuable concepts in all of science—and helps explain why some people are naturally so much happier than others and why an individual's happiness remains so stable despite seemingly large changes in external circumstance. Most physical systems and all healthy biological ones can maintain themselves in stable balance through feedback self-regulation. This stability isn't easy to achieve because internal and external conditions are always changing, threatening to shake things up. The result is a dynamic equilibrium: continuous short-term changes balancing out to create relatively uniform long-term conditions, despite all the temporary perturbations. Every cell in our body is a study in advanced homeostasis and their collective working together is a miracle of teamwork. Without a thought, we maintain a stable body temperature, blood sugar, heartbeat, blood pressure, oxygen level, acid–base balance, hydration, and a million other things. If we ate just one extra candy bar a day for a year (not at all hard for someone like me to contemplate doing), the weight gain would be about thirty pounds, and in a few years we would all look like sumo wrestlers. Something mysterious within us keeps us in weight balance.

Miraculously enough, humans are also able to maintain a homeostatic happiness balance, even when good and bad luck throws us marked ups and downs. In the mid-1980s, I did a study of men at high risk for HIV infection. They completed a battery of emotional assessments done just before and just after HIV testing and again six weeks later. In those early days, before any treatments were available, a diagnosis of AIDS was a death sentence, usually within a year and with the certainty of intense terminal suffering. Not surprisingly, the men who got the good news of a negative test showed a sudden, dramatic uplift in measures of emotional well-being. Equally not surprising, the men who got the fatal news of a positive test showed a sudden, dramatic plunge toward depression and anxiety. But very surprisingly, both groups were back close to their baseline happiness within six weeks. The best possible news and the worst possible news were equally unable to have a profound and permanent effect on positive and negative mood state.

Each of us has a kind of internal happiness thermostat that unconsciously regulates how we feel. There are constant variations around our average happiness depending on positive or negative life events, but the thermostat pulls us back to the mean—in just the same way as our automatic internal regulatory systems keep bodily temperature and weight in balance. "Hedonic adaptation" explains how the short-term shifts in happiness balance out. The thrill of winning the lottery doesn't last very long before the usual worries in your life intrude to bring you back down to earth. Being seriously injured in a car accident is devastating, but usually the devastation doesn't last forever. Getting what we want is never as wonderful as we hope; losing what we have is never as terrible as we fear.[7]

How happy we are is relatively stable over time and more determined by our happiness set point than by our income, age, health, marital status, and the number of friends we have. Most certainly, our happiness is not dependent on how much stuff we have, and it won't be hard for us to adjust to a future with a lot less stuff. Genetic studies suggest the setting of the happiness thermostat is about half inherited, half due to chance or early environment. The evolutionary advantages of an even-keel approach to life are obvious. Quick hedonic adaptation keeps us responsive to the present moment, but the happiness set point keeps us anchored in the long term. We can respond to the goods and bads in life without being overwhelmed with joy or crippled with despair. That is what homeostasis is all about. There is some variability among people in how close they stay to their happiness set point—some are more temporarily responsive to external circumstances than others—but sooner or later, most of us get back to an even keel.

Some people try to cheat the happiness set point with alcohol, substances, or psychiatric pills—but they find that homeostasis is tough to beat. Alcohol and drugs can give us temporary high points, but usually at the expense of at least equal and opposite low points. Psychiatric medications are helpful in recovering the happiness set point when it has been depressed by mental disorder, but there is no evidence it can raise the set point in normals.

There is no claim the set point is fixed forever or impermeable to change. Grinding poverty, long-term unemployment, divorce, death of a spouse, chronic illness, and serious injury can all lower the set point for extended periods, perhaps a lifetime. A transformational relationship, gaining wisdom about life, or therapeutic magic may raise it. But in most cases, we are what we are. Our economy can retool from growth to sustainability, produce lots less stuff, and still make far more than enough stuff to keep most of us happy.

The graph charting happiness against age over the course of an individual person's life cycle follows a fascinating U-shaped curve. People tend to have a happiness peak when they are in their twenties, then a midlife dip in their forties, and finally experience rising happiness as they age, with a second peak in the late sixties to early seventies. This finding is robust across genders, countries, times, and even species (chimps and orangutans also have their own age-adjusted midlife crises and late-life joy) and also holds up after controlling for wealth, marital status, children, employment, and other variables. The U-curve is even more pronounced in wealthier countries, where people live longer and enjoy a healthier old age. This patterning of happiness seems counterintuitive—who would have imagined old folks being at the top of the happiness game? But, on second thought, the finding makes sense. What makes us happy also shifts over the years. Blog posts written by younger people stress "excited, ecstatic, or elated," while the old favor "peaceful, relaxed, calm, or relieved." The pressure is off for us "oldsters"; our race has been run and we focus with less distraction on what really counts—appreciating little positive things in the present and remembering good moments from the past.[8]

There are two mystifying paradoxes in the relationship between gender and happiness. First, women have a slight edge over men on overall happiness, even though they are twice as likely to suffer from clinical depression and anxiety disorders. This discrepancy can be explained in two ways: either women in general have a broader emotional palette than men, or most women may be happier than men, with just a few exceptions who are very unhappy. Or it could be a little of both. Second paradox: subjective happiness in women has dropped consistently during the last four decades, even though objective measures of their well-being have consistently improved. This discrepancy probably reflects the increased stress of working and running a home and taking care of kids, amplified by the fact that more women are now sole breadwinners. Women would especially benefit from a growth-free economy, permitting them shorter workweeks.[9]

The many studies of how personality influences happiness consistently find that extroverts are happier, while those inclined to neuroticism (for example, guilt, anger, and anxiety) are unhappier. This is not exactly earth-shattering news and carries a strong whiff of tautology. Relationships make people happy and extroverts are better at forming them. Plus, it is so much easier to be extroverted when you are happy—happiness may cause the extroversion rather than the other way around. And it is hard to be happy when you are guilty, angry, and anxious. I think it might be more productive to see happiness as an independent personality trait that helps determine your other traits and behaviors. Indeed, a person's happiness set point may be one of the best predictors of how his life will turn out. If I could know just one detail before marrying or hiring someone, it would be how happy they have been over time. If I could have one wish for a child, it would be the blessing of a well-set happiness set point—much more important than being beautiful or smart. Being attractive plays only a small role in happiness for women and no role for men. And there is little relationship between being smart and being happy, once you control for all the other factors that go with intelligence.[10] Bottom line: a sustainable society can be a very happy society; an unsustainable society doesn't guarantee happiness in the short run and does guarantee misery in the long.

Gross National Happiness

Nothing in the predictors of happiness, in the scientific literature, in the sayings of wise people, or in common experience suggests that enduring happiness comes from having a billion dollars, or buying a mansion or your dream car, or winning the lottery, or eating in a many-starred restaurant, or owning a Picasso. You can't buy happiness in a store or order it on the Internet—no matter what the advertisements tell you. We have commercialized happiness and look for it in all the wrong places. Our genes are adapted to get maximum enjoyment from the very same things that were available to our ancestors as they wandered the world in small groups fifty thousand years ago. The best things in life are almost free—simple, spontaneous pleasures that satisfy and endure. Most of the rest is illusion and fleeting evanescence—the glitz of modern life is not gold.

We could improve our world, and increase its happiness by worrying less about Gross Domestic Product and working harder to improve Gross National Happiness. There is already considerable movement in this direction. The UN Sustainable Development Solutions Network began the "World Happiness Project" in 2011 to encourage participating countries to focus attention on happiness as a national goal for development. The project publishes an annual "World Happiness Report," derived mostly from the World Gallup Poll.[11] The Gross National Happiness (GNH) index is a much better indicator of long-term national success than the current standard, Gross Domestic Product.[12] GDP overemphasizes the benefits of economic activity and greatly underestimates the noneconomic activities that just happen to make people most happy—leisure, health, togetherness with family and friends, culture, education, exercise, safety, and nature. A country that races to produce and consume ever-more-useless things gets rewarded with a high GDP score, even though it is wasting resources, polluting the environment, and depriving its population of the time and peace to enjoy life.

Why not strive to increase the per capita happiness of a population rather than focus exclusively on its per capita economic production and consumption? The concept that governments should have as their highest goal the happiness of their citizens has a long history, starting with Confucius: "Good government obtains when those who are near are made happy, and those who are far off are attracted." Britain, France, Germany, Singapore, Thailand, and South Korea have shown particular interest in basing policy decisions not just on productivity and monetary reserves, but on creating "social reserves." Bhutan has traveled furthest in actualizing GNH with quantitative measures of economic wellness, environment, physical health, mental health, workplace wellness, social wellness, and governance. These are then aggregated and divided by population to determine "per capita happiness."[13]

What country you live in has a big impact on how happy you are. The most important predictors of national happiness are wealth, longevity, equity, free life choices, absence of corruption, and having people and a system to count on. National happiness also varies with time—during the past five years, happiness rose in sixty nations and worsened in forty-one. As global poverty is reduced, the world is getting somewhat happier and there is some convergence of happiness scores across different regions—like everything else, happiness is being globalized. The rankings at the top vary a bit, but Scandinavian countries, particularly Denmark, consistently lead the pack. Not hard to understand because they have vibrant economies, efficient governments, effective health care, a strong system of social safety nets, and less inequality in wealth distribution. It is somewhat surprising, though, that just one millennium after being bad-boy, berserking Vikings, the Nordics have become the world's best and most rational citizens—least likely to nurture societal delusions. They had once practiced the world's most aggressive brand of growth economy but have now become its best advertisement for rational sustainability. Other high scorers are the Netherlands, Switzerland, Austria, Canada, Australia, and New Zealand. Latin American countries have made the biggest recent gains and play well above their economic weight, almost certainly because of the paramount value they place on close family and community ties. The worst places to be are the greatly overpopulated, war- and terror-plagued Middle East, most of Africa, and Afghanistan. Sustained happiness depends on stability and security.[14]

Despite its wealth and power, the United States usually finishes somewhere between fifteenth and twentieth on the happiness hit parade. And we aren't getting happier as we get richer. Despite an astounding increase in GDP and access to stuff, our happiness ratings are stuck today just about where they were four decades ago. Because of increasing wealth disparity, the rich have more money than they know what to do with, while average people are just getting by, and the poor are desperate. We would be a much happier country in aggregate if we spread our wealth more equitably and provided social supports for those dipping into poverty. The very rich need much less than they have; the very poor need to get to the basic minimum. Small-percentage income gains would make the poor and middle class much happier than the same percentage increase does for the affluent. Because our economic and tax policies allow the rich to get much richer at the expense of everyone else (reducing our average national happiness), the fastest way to be a happier country would be to tax great wealth and conspicuous consumption and use the proceeds to fund social services and sustainability projects. Less greed is good; more all for one, and one for all.

People don't get to be billionaires by worrying about the happiness of others. Trump is among the most greedy and grasping of people and has proposed a reverse Robin Hood tax program to help himself and his buddies take an even bigger slice of the pie. His billionaire cabinet turns Lincoln's "government of the people, by the people, for the people" into government of the superrich, by the superrich, and for the superrich. Of course, Trump is also the world's icon of conspicuous consumption, both in his own life and in his life's work of seducing people to buy vulgarly extravagant (and totally tasteless) things they don't need. Meanwhile, for the world's less privileged—"let them eat cake."

Finding Happiness in All the Right Places

*If you need sunshine to bring you happiness, then
you haven't tried dancing in the rain.*
—UNKNOWN

The things that made us happiest fifty thousand years ago are still what make us happiest today. The millennia pass and succeeding generations of mankind create new and fancy frills to pass the time, but the basic ingredients of a good life remain remarkably the same over the ages: family, friends, finding meaning, feeling gratitude, giving to others, loving nature, staying fit and healthy, and a spiritual appreciation of the implausible wonder of life. Have these and it doesn't really matter how much stuff you have; don't have these and being a billionaire won't save you. Happiness surveys converge on the important point that we look for happiness in all the wrong places. Happiness exists under our noses and in the everyday simple things. It comes from being in harmony with nature and with our human nature. It doesn't come from being a cog in a growth economy clogged with shopping malls and Trump hotels.

George Herbert, an otherwise undistinguished contemporary of Shakespeare, wrote only one line that endured the test of time: "Living well is the best revenge."[15] Too often in the past, this has been used to justify just the kind of gilded Trumpian lifestyles that our society can no longer afford to indulge. Two hundred years ago, Edmund Burke warned: "If we command our wealth, we shall be rich and free. If our wealth commands us, we are poor indeed."[16] Wealth doesn't buy happiness, but poverty does buy misery. And wealth can also buy misery, when (as often happens) it becomes an end in itself and a consuming preoccupation. Happiness correlates closely with income only up to about $75,000 per capita per year (or its equivalent in other countries). After reaching that level, making more money doesn't make you much happier.[17] You start running on a tireless, but tiresome, treadmill—the more you have, the more you think you need, and your comparison group in judging success gets tougher and tougher to compete with. Once essential needs are met, the best things in life are priceless and can't be bought. Among the unhappiest people I have known and treated were the richest. One of the happiest is my best friend in San Diego, who has been involuntarily homeless for twenty years, but nonetheless is still "having the time of his life" hanging out in libraries and bookstores, learning all there is to know about our world.

Because money was invented so very recently, evolution has not equipped us with a healthy homeostatic mechanism to control our desire for it. We lack the usual satiation effect that keeps us from eating out the whole candy store—with money, the more you have, the more you still want or feel you need.[18] Money distracts from the simple pleasures that are more pleasurable. The superrich people I have known spend so much time and worry building and constantly repairing their palatial mansions, they never get to live in and enjoy them. Howard Hughes, an expert on the special misery that can accompany being a billionaire, ruefully summarized his life experience in a not particularly original, but nonetheless completely accurate, way: "Money can't buy happiness." The good news is that we won't suffer a happiness crash if our GDP declines in a steady and gradual way—exactly what it needs to do if we are to achieve worldwide sustainability. Living well in the future will have to mean living smaller, smarter, less wastefully. Simple and happy are usually the same. Trump is one of the richest and most powerful men in the world, but he seems to be one of the unhappiest. Nothing he has and nothing he achieves ever seems enough to make him feel at all comfortable in his own skin.

Psychologists conducting surveys of "subjective well-being" distinguish two types of happiness—moment-to-moment pleasure, and long-term life satisfaction.[19] The best example of how different these can be comes from mothers. Being with kids is not always that gratifying, particularly if childrearing is squeezed into a busy workweek or there are lots of, or difficult, kids to rear. But that doesn't mean they regret being mothers—having children is, in the long term, a very satisfying experience that is usually worth the temporary short-term sacrifices and discomforts.

If we are to avoid succumbing to societal delusions, we must also take the long view. The sacrifices we make now to protect the welfare of future generations will inevitably require reducing our expectations for immediate hedonic pleasure in ours. But there is a deep satisfaction in doing right for our children and grandchildren and in leaving the world a better, not a much worse, place than we found it. We all experience this feeling of fulfillment in doing the right things for members of our own families. We must now generalize it to doing the right thing for the long-term future of our species, other species, and the planet we share.

Being a Mammal

Man is a mammal, defined most fundamentally by our ability to love, to attach, and be loved by others. First (at mother's breast), last (saying last good-byes to those we have loved), and always— our most meaningful moments are shared moments. By far the most consistent finding from decades of research is that happy people are connected people. Man cannot ever be fully happy as an island. But relationships take time we no longer have, when most of our waking lives are spent working, commuting to work, and looking at screens. If we produce and consume less, we can have the time to enjoy life more. And a big part of our desperate need to acquire things arises from our having lost the much deeper pleasure of having people.

Relationships come up tops in every list of what makes people happy. Having a strong social network doesn't guarantee great happiness, but it is a prerequisite for it. The happiest 10 percent of people are those who spend the most time with those they love.[20] Chasing the fool's gold of stuff switches meaningful attachments from people to things, and thus makes them less meaningful. Marcuse's shortened workweek would make us less wealthy in material possessions, but much wealthier in our ties to family and friends.

Of course, not all relationships are equally important—it is the quality, not the quantity, that counts. One or two or three very close friendships or family relationships may be enough; having a large network of close relationships is nice, but much less essential. Most crucial is how comfortable we are in disclosing ourselves to others and sharing their self-disclosures to us. The closest relationships are the most open ones and the ones that promote sharing and cooperation. It's not clear what is cause and what is effect, but for all its problems, being married is strongly associated with being happy. Part of this may relate to married couples often being best friends and also having better sex. Contrary to Hollywood fantasy, married people have more sex and enjoy it more than singles—familiarity and availability have their charms. And, no great surprise, having a good sex life is very strongly correlated with happiness—evolution would have it no other way.[21] But, for some people, friendship brings more happiness than family—as George Burns put it, "Happiness is having a large, loving, caring, close-knit family living in another city."

Meaning, Acceptance, and Gratitude

Aristotle presented the first and best philosophical analysis of human happiness. He asks, "What is the ultimate purpose of human existence?" He answers: "Happiness is the meaning and the purpose of life, the whole aim and end of human existence." Aristotle's definition of happiness (*eudaimonia*) includes our fleeting pleasures, but far transcends them. He saw purpose everywhere in the world and man's natural purpose is to virtuously live up to our full potential. "One swallow does not make a summer, neither does one fine day; similarly, one day or brief time of happiness does not make a person entirely happy." For Aristotle, being happy meant a lifetime devoted to doing good.[22] Good advice then, still good advice now. Aristotle would of course be appalled by Trump, but he would also despair of our societal proclivity for fleeting pleasures at the expense of rational thought and grown-up responsibility.

Finding deep meaningfulness in life always requires sacrificing some of its superficial pleasures. Raising kids means trading in your pleasure for theirs, but most of us eagerly (perhaps too eagerly) make the trade because the long-term satisfactions are more significant than the short-term sacrifices. The happiest people I have known are those who spend their lives helping others, selflessly doing work they deeply believe in. The day-to-day grind for teachers, nurses, social workers, therapists, nuns, journalists, NGO workers, volunteers, and other do-gooders is harried, stressful, overworked, and underpaid. But their do-gooding can be so underpaid in dollars precisely because its payoff is so great in personal satisfaction. The unhappiest people I have known are those who feel their life has lost all meaning—parents whose children have died, workers who retire from or lose their jobs, the very rich who have nothing to live for, the very poor who have nothing to live on, the lonely.

Accepting life gladly, and on its own terms, is a daily challenge to all of us and a central theme of most religions. My favorite Christian acceptance quote, from theologian Reinhold Niebuhr, has become the motto for Alcoholics Anonymous: "God, grant me the serenity to accept the things I cannot change, the courage to change the things I can, and the wisdom to know the difference." Taoism speaks of "wei wu wei": literally "effortless doing"; loosely translated into American basketball lingo as "let the game come to you," doing what is natural to the time and place rather than striving for the impossible. Confucius offered: "Everything has beauty but not everyone sees it." My favorite from Judaism: "There is a time for everything, and a season for every activity under the heavens: a time to be born and a time to die." Mahatma Gandhi, a Hindu, famously observed, "An eye for an eye only winds up making the whole world blind." My favorite from Buddhism comes from the Dalai Lama: "I do not judge the universe."

Grateful people tend to lack envy, be generous, not depend so much on material goods, and deal more gracefully with the aches and pains of everyday life. It is my clinical experience that successful psychotherapy is furthered by, and results in, a dramatic increase in the person's gratitude and appreciation for all the little things that make life wonderful, or at least bearable. All of us have lots to be thankful for, and not just on Thanksgiving Day.

"It is more blessed to give than to receive" (Acts 20:35). Sounds corny, but it turns out to be empirically true. The cynical view of human nature is that we are selfish takers. Experience confirms this to be true of all of us some of the time and some of us all of the time (think Trump). But we also have within our instinctual hard wiring a genetic apparatus for altruism that exerts a powerful influence on our social behaviors and our satisfactions.

Controversy surrounds the survival value of sharing—why is it that so many nice guys have not finished last in the evolutionary race? One school argues that it is a smart play for my genes to promote giving to, and sacrificing for, my close kin because they carry a similar genetic package. My selfish DNA benefits by making me unselfish—it will wind up ahead in the game of survival if I choose to die so that three of my siblings might live, even though I personally wind up on the losing end. Another less popular but still plausible school of evolutionary thought argues that altruism survived at the group level—tribes that were especially good at promoting internal cooperation were more likely to thrive than those that weren't. Altruism genes probably made the cut both ways through improved individual as well as group survival.[23] Either way, they make us better people and are the great hope for our species.

It may seem counterintuitive that giving generates more pleasure than receiving—when asked, most of us think we will be happier on the receiving end of a gift. But research consistently confirms the heartwarming "hedonistic paradox" that we are not just a one-dimensional, greedy, self-serving "*homo economicus*." Across all sorts of experimental paradigms, people report feeling greater happiness in giving, rather than in receiving. We are blessed with empathy for other people, which allows us to take enjoyment in their pleasure. How nice to know that we are happiest when we make someone else happy—genetically programmed to be generous and good, not just selfish or evil.[24]

When we are generous to others, they are generous to us—creating a benign cycle of reciprocity that cements mutual bonds and societal cohesion. Social exchange encourages alliance and promotes sharing the wealth during tough times. The classic example from anthropology was the potlatch parties held by Native Americans in the Pacific Northwest—highly ritualized commemorations of births, weddings, deaths, victories, and other special moments in the life of the individual and the tribe. The most successful men would work like crazy all year to accumulate vast stores of goods, in order to then give them all away, in an orgy of ostentatious generosity, to those less fortunate or less enterprising. The "big man" was rewarded with prestigious political and religious titles and increased power in tribal decision making. The group was rewarded by having a redistributive system that guaranteed everyone made it through the winter.[25] And a war of competing generosity beats the hell out of establishing prestige with spears and fists. You get to lead only if you are best at bringing in the bacon—and at giving it away.

We now have a destructive reverse potlatch system that disadvantages the hordes of the earth so that the few big men can become even bigger (think Trump and his cabinet of expropriative exploiters). Fortunately, there are other big men (think Bill Gates, Warren Buffett, George Soros, Michael Bloomberg, Mark Zuckerberg) who are using their billions to better the earth. Given our ridiculous inequality, we must hope that more of the other 80 superbillionaires (who own more wealth than the world's 3.5

billion poorest people) will follow their example. Once basic needs are met, money functions increasingly as an abstraction or a means to obtain status and power. As it stands now, too much concentrated money is competing to chase too many vastly over-priced, really dumb things. You know our value system is totally screwed up when one of Jeff Koons's banal concoctions sells for $60 million, while the serious work of improving the world goes so badly unfunded.

We also need to enlist everyone's better angel, not just the big bucks of the superrich, in a concerted effort to share, not grab, the earth's diminishing bounty. Public policy should promote volun-teerism, community involvement, and citizenship. A fair system of inheritance taxation would help prod those less sensitive to the nonmonetary incentives. And giving can't just be local. We must feel kinship across the planet because a problem in Timbuktu can quickly become a problem on Times Square. Prestige should go not just to the accumulation of wealth, but to its just redistribu-tion; not to the exploitation of others, but to their elevation. We shouldn't glorify *Forbes* magazine's richest four hundred people, but rather extol the most generous and wisest givers. "Big man" satisfaction should come from contributing to the welfare of all. And all of us, in our own small ways, can be "big men."

The best and most natural way to go from misery to happiness is to find a cause you believe in and to work hard for it. Many of my friends are experiencing an unexpected and newfound patriot-ism and satisfaction working in the growing populist movement to defend democracy against Trump's incursions. My hope is that, in some small way, this book will encourage others to follow suit.

Caring for the Body

"The greatest wealth is health." Virgil said it all two thousand years ago. Health and happiness are a two-way street. Being healthy is not necessary or sufficient to being happy, but it is the best possible head start. And being happy is a powerful predictor of longevity—with an impact as large as smoking status. Despite hype to the contrary, it doesn't appear that happiness cures the sick or extends their lives, but it does seem to protect the healthy. From a public health perspective, we would be a healthier country if we wasted less money on excessive (and too often harmful) medical treatments and instead spent more money on social programs that helped people be happier and feel more secure. Improving the Gross National Happiness index is certainly a better means for improving health than pushing for a higher GDP.

Exercise is good for the heart, the mind, the spirit, and the waistline. It improves physical and mental health and is the only proven way to ward off dementia. Physical activity has always been an essential part of being human, but is no more. Our genes, preadapted to the average expectable environment of fifty thousand years ago, expect us to exert ourselves hard every day to obtain relatively meager supplies of food. Our Paleolithic ancestors were in great shape. They had to be. Survival required a daily round of major physical activity: hunting, scavenging, and gathering food; preparing and cooking it; collecting firewood and building fires; drawing and carrying water; building and maintaining shelters; the mating game; child care; seasonal migrations; war and defense; and communal song and dance. The subsequent agricultural and industrial revolutions changed the nature of physical labor, and reduced its diversity, but still demanded and produced a population that was extremely fit.

This all changed when recently we substituted fossil fuel for muscle power. Forty percent of us are now couch potatoes and most of the rest are much more sedentary than nature and our genes ever intended. Exercise, once a necessity, has become a luxury. Our genes are not prepared to deal with an environment that requires almost no physical activity and provides well-stocked refrigerators, fast-food joints, and superhuman portions. Inactivity and overeating are a terrible combination. The fatal paradox is that while our need for calories has diminished greatly, our intake of calories has simultaneously exploded. The result is a massive collective reduction of muscle mass and the massive collective accumulation of fat stores.

We are in the midst of an obesity epidemic that has supplanted smoking as public health enemy number one. Two-thirds of adults and one-third of kids are overweight; one-third of us are obese. Everything has had to be resized to accommodate our ever-bulkier bodies—clothing, plane seats, even coffins. Being obese is the major risk factor for many lethal medical conditions and a frequent cause of unhappiness, stigma, and low self-esteem. Those groups that experienced the most scarcity in the past—for example, Polynesians and Native American Pimas—are most energy efficient in the present, best able to conserve calories as fat, and therefore most likely to become obese and suffer its discomforts and diseases.[26] The United States leads the world in obesity, but other countries are fast catching up—even in the developing world. Globalization of food supplies and tastes is resulting in deteriorating diets and consequent health problems everywhere.

Our government policies are making things worse. We should be subsidizing healthy, low-calorie foods. Instead, the powerful agribusiness lobby has our government pouring billions into corn-derived fructose sweeteners that tempt us to eat all the wrong things. Commercial greed makes it cheap to eat badly, expensive to eat well. A big salad should cost one-third of what a Big Mac costs, a small Coke a small fraction the cost of a large Coke, and supersize portions should be eliminated. I don't believe the arguments justifying the status quo—that being obese is a civil right, that we should support fat diversity, that fat is beautiful, that being heavy is what "real" people look like, or that obesity is simply a genetic disease. To contain obesity, we need to change societal policy and individual behavior. If this seems impossible, think about the health miracle that occurred when public outrage forced pusillanimous politicians to finally take on

Big Tobacco. Smoking was cut by two-thirds. A similar public health effort could have a similar positive impact on obesity. If we want to be healthier and happier, we need policies that encourage much more exercise and discourage caloric overdoses.

As it was in ancient Greece, where sound body partnered with sound mind, physical activity should be an integral part of education from preschool through graduate school. We shouldn't be spending trillions of dollars on often unnecessary medical care and should be spending billions of dollars subsidizing a gym membership for every citizen. We shouldn't have drug and food ads on television and should have exercise and healthy eating ads. We have lost our vital and immediate connection with our bodies and need to regain it. For some, this will mean formal workouts. For others, it will simply mean walking to work and climbing the stairs, instead of driving and taking the elevator.

It will be hard to get couch potatoes started, but once off the couch a virtuous cycle kicks in. Exercise releases endorphins that provide a pleasurable sense of satisfaction. The more you exercise, the more you will want to exercise to re-create that feeling. The more connected to your body you feel, the more you want to use it and feed it well. I know the drill—I fight a daily battle against weight gain, always losing it, but not by much.

Feeding the Spirit

The vast universe may be purposeless, but on a human scale we must all find personal meaning in it. I am not a religious man in any conventional sense, but I have great respect for the benefits others derive from religious faith and experience. Religion provides meaning in a world that can seem random and meaningless; community in a world that can be lonely; hope when things seem hopeless; consolation in the face of loss; and courage when confronting danger and death. Religious feeling is so ubiquitous; the potential for it must be hardwired in our brains. In the tough, do-or-die world of our ancestors, great survival value would accrue to those who could face exigencies buoyed by religious belief and a community of fellow believers.

Surveys by Gallup, Pew, and the National Opinion Research Center find that religious people in the United States are happier, have more life satisfaction, better sex lives, and lower rates of drug abuse, depression, and suicide than nonreligious people. Of course, this correlation does not prove causation. It could be that unhappier people, or those who have suffered more, become disillusioned and leave their religion. And the unhappiness among the nonreligious could result from their being a minority group in a religious country. The Netherlands and Denmark are two of the happiest countries on earth—as well as the least religious.

Organized religions also have their limitations and can create problems. Many of us find the stories and beliefs of the world's religions poetically beautiful, but impossibly implausible. In such an enormous universe, it doesn't seem remotely likely that "not a single sparrow can fall to the ground without your Father knowing it" (Matthew 10:29). It's much more likely we created God in our image than that he created us in his. And the belief that my God is better than yours is both naïve and dangerous. There are no chosen people except in the self-serving eye of the beholder. We are all God's children or none of us are. That religion is a necessary precondition to morality is disproved by the frequency of moral atheists and immoral religious hypocrites. It would be a better world if only we could bridge the divisive religious discord that leads to war, and end the fecundity competition among the orthodox that leads to overpopulation. Organized religions can help solve our societal delusions or make them intractable—mostly depending on how strongly they cling to the fundamentalist dogma of the past as opposed to responding responsibly to the needs of the present and the demands of the future.

If orthodox religion is a mixed bag of blessings and curses, spirituality in all its many forms is almost always a benefit, and carries few risks. Everyday life has its everyday pleasures, but also its inevitable banalities and disappointments. Spiritual experiences allow us to live fully and richly—achieving tranquillity, peace of mind, acceptance, and improved physical and mental health. It is easy to get lost in the forest of quotidian experience and lose track of the weird wonder of our existence. Peak happiness often occurs when we immerse ourselves in each moment as it passes, detached from past regrets or future worries and expectations.

There are hundreds, perhaps thousands, of schools of formal spiritual and mindfulness practice, but I prefer my personal route of simply living in the moment. My spirituality is getting lost in nature, thrilling at the implausibility of being alive, reading a great line in *Ulysses,* hearing a song, chocolate mint ice cream, a grandchild's smile, and a thousand other things that thrill me and make my spirit soar. Each person must find his own source of spiritual nourishment, but life is certainly dull and cardboard without some portal to at least temporary transcendence. Happy people are usually spiritual people and spiritual people are usually happy.

Pursuing Happiness in Just the Wrong Place: The Substance Trap

Marx said that religion was the opium of the people—blinding them to worldly problems and encouraging passivity in accepting the unacceptable status quo. Nowadays, psychoactive drugs are the feel-good facilitators of societal denial. Almost one-third of adults in the United States take a legal or illegal drug with the purpose of reducing psychic or physical pain. And so many people take a whole cocktail of drugs that drug overdose has become a leading cause of death.[27]

In *Brave New World,* Huxley called his all-purpose, feel-good drug "Soma," borrowing the name from religious works written in Sanskrit two thousand five hundred years ago.[28] Soma was a god, and also a ritual drink, praised in hundreds of hymns for its stimulating behavioral, medicinal, and spiritual effects. The kicker in Soma was probably ephedra, a chemical still in use today in medicines, as a performance enhancer, and as the raw material for making methamphetamine. Huxley's attitude toward psychoactive drugs was decidedly ambivalent and changed dramatically as, over his lifetime, he enlarged his personal experiences with them. In *Brave New World,* published in 1932, Soma is a dangerous seducer—narcotizing our souls, making us less than fully human. Twenty-six years later, in "Drugs That Shape Men's Minds," substances are useful tools to help us find our souls and sharpen our perceptions.[29] Enough good trips can turn even the most skeptical abstainers into the strongest believers.

Huxley follows in a long tradition of drug devotees. Evidence from archaeology and anthropology shows that humans have been getting high as long as there have been humans. And animals were doing it long before. Addiction turns out to be a cross-species vulnerability, not something invented by us. Our primate ancestors also liked their drink whenever Mother Nature was willing to serve as bartender. Uneaten fallen fruit ferments to produce alcohol, thereby offering its consumers an unbeatable combination of concentrated calories and a nice buzz. Vervet monkeys in the wild display patterns of substance use that closely mimic our own. Adolescent vervets get drunk more often than adults and adults can be divided into teetotalers, social drinkers, binge drinkers, and end-stage alcoholics, in proportions similar to what we see in humans.[30] Perhaps God ordered Adam and Eve not to eat forbidden fruit for fear they might get hooked on it. How long was that apple lying around? Was the snake the first drug pusher?

Animals in the wild also abuse a whole variety of naturally occurring psychoactive substances, which are produced by plants to protect themselves against parasites and browsers. We like them too. Opium comes from poppy plants, marijuana from cannabis plants, cocaine from coca plants, psilocybin from mushrooms, nicotine from tobacco, caffeine from coffee, amphetamines from khat. Horses like locoweed, felines like catnip, jaguars like a psychoactive jungle vine, reindeer like mushrooms, wallabies like opium poppies, and pigs like cannabinoid-containing truffles. Some animals produce defensive poisons that have their own psychedelic charms. Monkeys and lemurs get high off the toxic chemicals secreted by millipedes. Some bird species do a poisonous ant rubdown—probably to gain protection from parasites, but also to get high. Poisonous toads can be milked monthly to provide a psychoactive juice that has commercial value. Nature is a mind-bending drugstore.[31]

Human ingenuity has built on nature to produce ever-stronger drugs. Our synapses were evolved to balance the effects of a hundred or more neurotransmitters, each acting as a team player in homeostatic harmony. Modern drugs take over the orchestra pit and totally overwhelm the rest of the band. Cocaine and the amphetamines block the clearing of dopamine from the synapse, skyrocketing the reward system to places it was never meant to go—and then causing inevitable crashes and cravings when the dopamine surge is removed. The invading external drug totally dominates and enslaves the pleasure reward system. Starving rats will pick cocaine over food (as will starving human cocaine addicts) and die of exhaustion, pressing a lever thousands of times an hour for days just to keep their cocaine fix coming. Rats will also cross a painful electrical grid to get their fix—just as withdrawing cocaine addicts will do the most extreme things to get their magic powder. Nicotine and caffeine have much less profound effects on dopamine, but they are still powerful enough that hundreds of millions of people are addicted to them.

In a parallel and even more dangerous and potentially deadly way, heroin and the prescription opioid narcotics overwhelm the pleasure system's internal endorphins. The receptor sites are saturated, massively amplifying a usually tame and useful system, and creating an experientially irresistible craving for more. The cortex may want to stop using the drug, but loses the argument to the hungry receptors.[32]

Alcohol works its wonders and harms because it affects so many aspects of neural transmission. It has a high and rapid affinity for neuronal membranes, inhibiting the normal flow of ions in their channels; it impacts the work of enzymes and it binds to receptors for acetylcholine, serotonin, GABA, and NMDA. In moderation, alcohol is harmless, perhaps even helpful, to health. But there is a big downside. About 8 percent of the population becomes addicted to alcohol in ways that can result in short-term behavioral deterioration and long-term dementia and liver disease. Alcohol is a major contributor to death from accidents, homicides, suicides, and disease.[33]

The United States is facing its worst-ever epidemic of opioid addiction, now also spreading all over the world. More than thirty thousand people die every year and many millions are iatrogenically hooked. Poppies have been used forever for medicinal, spiritual, and recreational purposes—always causing some damage, but nothing like the current carnage. Pharmaceutical industry pill pushing is most responsible, along with the synthesis of ever-more-potent opioid derivatives (carfentanyl is ten thousand times more powerful than morphine). Underlying the careless prescription of pain medicine is the widespread expectation, among doctors and patients, that there is a pill for every problem, a quick fix for every pain and distress. Individually, and as a society, our quest for simple solutions to complex problems often makes them much worse.[34] There has never been a time in the history of the world when buying street drugs was so dangerous—so many are secretly laced with superpotent, deadly synthetics. But this doesn't stop the drug trade—legal and illegal—because its profits are huge, its audience captive.

Pot also mimics naturally occurring brain neurotransmitters. It has its own set of dangers, but nothing compared to the havoc wreaked by the opioids. In a rational world, pot would be legal and prescription opioids illegal, not the other way around—but drug companies are much better at political lobbying than drug cartels. States that legalize pot experience a quick and dramatic reduction in opioid overdose deaths.[35]

Very little happens in the brain without serotonin. Its fourteen different receptor types help regulate the workings of all the other neurotransmitters to modulate mood, anxiety, aggression, sexual function, appetite, sleep, learning, memory, nausea, and temperature regulation. Serotonin had its humble beginnings long ago in the earthworm but is now the target of the most popular drugs used in psychiatry, and some used in medicine (for nausea and migraine). The selective serotonin reuptake inhibitors (SSRIs) are useful when used by the relatively few who really need them, useless or harmful when overused by the many who are seduced by drug company marketing. SSRIs are effective only in treating symptoms of mental disorder, not possessing Soma's wondrous qualities as happiness pills. And Soma had no side effects and caused no withdrawal; SSRIs produce both. For most people taking them, SSRIs offer no more than a very expensive placebo with little benefit and significant risk.[36]

The United States is especially addicted to drugs. We have 5 percent of the world's population but consume 50 percent of all prescription meds sold in the world and 80 percent of prescription opioid painkillers. Ten percent of Americans have used an illegal drug in the past month—that's twenty million people smoking pot, five million on another substance. Ironically, alcohol is the only substance used in relative moderation in the United States—we rank forty-eighth among countries in per capita yearly alcohol consumption. Some people desperately need the meds they are taking to treat severe psychiatric problems or pain. And for the majority who can participate in moderation, recreational substance use is mostly harmless fun. But it is quite destructive for many, and even deadly for some. And our almost ubiquitous use of substances risks turning us into a sick society. Expecting quick chemical fixes generalizes easily into expecting quick political fixes. It is hard to have a grown-up, resilient, and responsible citizenry when a third of the citizens require a pill or a drink to get through the day. Our pill-popping society facilitates our societal delusions and societal deluders. We must gradually withdraw from our unprecedented substance dependency if we are to become a mature society that can shed its denial and face reality as it is.[37]

What Counts

When you have lived a long life, you learn what counts and what doesn't. The things that make me happy are immediate, mostly almost free, and easily accessible. Feeling the sun on my skin and the wind in my hair. Summarizing the morning paper to my wife. Discussing history with my older grandkids and playing basketball with the young twins. Watching old movies. Learning new facts. Rereading the books I have loved. The deeply satisfying pleasure of an afternoon nap on the beach. Visiting a strange new place or revisiting a beloved familiar one. Pizza. Seltzer water. A Snickers bar. The fact I can still walk and swim and hit a tennis ball. A hug, a joke, a family meal, a puppy, an orange sunset, the satisfaction of having done the right thing, the giggle of a child, a pretty turn of phrase or shape of ankle. Being only halfway demented. Making love. My ambition now is never again to purchase anything I can't consume within a few weeks. I hope to live out my life with no new clothes, furnishings, or tchotchkes. Cars and gadgets will be used till they die—hopefully after I do. No new entangling alliances with material objects. Things don't make you happy. People do make you happy. With a little luck, I will hold on to the people I love, help them along when I can, and not become too much of a burden.

We are living at a time of unprecedented plenty, long life, personal security, relative peace, low crime, clear air, clean water, and technological marvels. Compared to historical norms, this is the best of times for most people in developed countries. It is relatively easy to feel satisfied when the sailing is smooth, but none of us can expect happiness to mean a future free of sacrifice. All of us must learn to live less for possessions, more for people.

We have no reason to despair and no excuse for feeling sorry for ourselves. It has never been easy being human and our current challenges, difficult as they may seem, pale in comparison to the black plague or the Thirty Years' War or biblical droughts. Our civilization, and perhaps species, can survive only by downsizing selfish expectations and upsizing altruistic cooperation. This will require a major shift in our behaviors and institutions—but it is comforting to know that our future is completely in our hands, to preserve or destroy.

Societal delusions protect things that are not worth saving, nonessential to real happiness and well-being. The human species still has an excellent hand to play, if only we wise up soon enough and use our cards well. We humans have a long tradition of resilience when confronted by crisis—we have repeatedly risen to challenges and certainly have the strength within us to deal gracefully and effectively with the sacrifices now required of us. I still have faith that we can not only endure Trump but prevail against him—and the societal delusions he represents.

Chapter 9

Team Earth

We may have all come on different ships,
but we're in the same boat now.
—MARTIN LUTHER KING JR.

No Man Is an Island

We Americans have something of a split personality—part competitive lone wolf, part cooperating member of a well-organized wolf pack. The lone wolves are best embodied by the movie actor John Wayne. "The Duke" played himself in 169 movies, spanning fifty years—always big, brash, tough, self-sufficient, self-contained, making his own way, needing no one. America's most iconic hero—but not its most lovable. And not really the most accurate representation of who we are. At Christmas every year, my family and I (and probably most people) much prefer watching another more approachable and appealing portrait of the American way. Jimmy Stewart's *It's a Wonderful Life* is a bighearted celebration of public-spirited community-mindedness, of good neighborliness, and of the joys of human interdependency. The happy ending is a triumph of and for the whole town—not a reward for one man's independent struggle and achievement.

Individuality has a long arc in American consciousness—it is central to our foundational myth and persists as the main sell of recent political propaganda. We revere the original colonists as liberty lovers, making their fortune and practicing their faith in a New World, free from the external constraints of the old. Hollywood westerns extend the metaphor to the lonely cowboy, who relied only on his wit, grit, and gun to face down bad guys, "Injuns," and hostile nature. And then came the politicians. Herbert Hoover coined the term "rugged individualism" first to help win the election of 1928, and then to explain his passivity in response to the country's sufferings during the Great Depression: "We were challenged with a choice between the American system of rugged individualism and a European philosophy of diametrically opposed paternalism and state socialism. The acceptance of these ideas would have meant the destruction of self-government through centralization of government." Hoover believed that government help would ruin "the initiative and enterprise of the American people." He was dead wrong—his rugged Republican individualism was the worst possible economic and human response to the Depression, making it much crueler than it had to be. FDR's New Deal created jobs, helped turn the economy around, and softened the blow for people who had no other source of support. Trump and the Kochs are the modern Hoovers, trying to undo the protections for average people that were built into the New Deal (and have been extended by Democratic presidents ever since). The United States already provides the most meager social and economic safety net of any developed country; the GOP would like to cut it altogether in order to give the rich yet another tax break. This is what the Trumpcare battle is all about.

The GOP is as wrong now as Hoover was then. American life has always been much more grounded in cooperation than competition. The early colonists lived in the tightest of communities—survival was virtually impossible outside the group and intolerable without its approval. And, unlike the movies, the real old West was civilized, collaborative, and much less violent than many modern big cities (unless you happened to be a Native American, targeted for removal or extermination by the U.S. Army). Regulation governed every aspect of daily life. Wagon trains agreed upon constitutions before they headed west. Mining towns had tight rules governing claims and digging rights. Ranchers and homesteaders created land associations to settle title and boundary disputes. Gun control was much stricter in the old West than it is in the new—you checked your hardware with the sheriff before you had the freedom of the town. Married women were strictly off-limits to roughnecks. People went to church on Sundays. There wasn't much tolerance for loners or outlaws or gunslingers. The justice of the peace interpreted the law and the sheriff and his posse enforced it.

Community responsibility is, and always has been, as American as apple pie. Cooperation expresses our decency and is essential to our success. The pioneer tradition was to help your neighbor today and expect that he will help you tomorrow. Communities joined together cheerfully for house raisings, barn raisings, and church raisings. Everyone understood that luck played as large a role in life as effort and endowment. Sharing provided an insurance policy against adversity and bad luck, spreading the risk and burden throughout the group. And it is a myth to think that our ancestors came to America as rugged individualists, each self-reliantly making his own way. Most often (as in my family) one person came first, saved some money, and gradually brought over his brothers, sisters, parents, and extended family. Everyone shared and felt responsible for everyone else.

"E Pluribus Unum" became the motto of the United States on the very first Independence Day, July 4, 1776. "Out of many, one"—a paraphrase of Cicero: "When each person loves the other as much as himself, it makes one out of many." Two hundred and twenty-eight years later, Obama raised the same inspiring banner: "There is not a liberal America and a conservative America—there is the United States of America. There is not a black America and a white and a Latino America and Asian America—there's the United States of America. . . . There are no Red States and Blue States, there is the United States of America." It's now the people's job to counter Trump's divisiveness and make America whole again. We certainly can't trust our politicians to bring us together—too many are enslaved by the financial and ideological conflicts of interest that drive us apart. Most pols will serve the public interest only when they become more afraid of the public than of the special interests that now bully and direct them. Women would not have the vote, blacks and gays would not have civil rights, and our environment would be in even worse shape were it not for the will of the people gradually shaping the agenda of the politicians. The powerful upsurge of people power opposing Trump is our best bet for recovery from our current political insanity.

Progressive populism has succeeded because people were willing to struggle for something they strongly believed in. The most meaningful happiness comes from working for a better world and for our children. Many people despair at living in our Trump dystopia—on the left, feeling defeated and hopeless about the future; on the right, many of Trump's supporters feel betrayed by him and impotent to change their lives for the better. Despair is a losing game, both for the individual and for society. Americans come together in times of stress. It feels good and it gets things done. In Chapter 6, we discussed a "We the People Contract" that could gain majority support and meet many of the needs of both the left and right. There is no better way to "Make America Great Again" than to engage people for the common good and to stop all the damned divisiveness. We must stand up for right and back down on might, in all the places it is being misapplied.

I take heart in the achievements, small and large, of past populism. Here are some of my favorite examples. They inspire me and I hope they will also inspire you. Every mighty river starts with just a few small drops of rain.

Dog Poop

This is a humble, but telling, tale of a bottom-up social change that in its own small way shook the world and made it a cleaner, nicer, more civilized place to live. It all started forty years ago, when citizens of a small town outside New York City were outraged by the unfettered and formidable fecal productions of one very large Great Dane and successfully banded together to ban the town's dog owners from dirtying their streets. This one small initiative in a perfectly obscure place set off a completely unexpected chain reaction, worldwide in its scope and permanent in its duration. The new social movement soon spread across the Hudson River to the Big Apple, in those days home to half a million dogs and one hundred tons per day of dog doo. After some controversy and political infighting, New York City managed to pass the world's first Canine Waste Law, with the delicately worded injunction, "A person who owns, possesses or controls a dog, cat or other animal shall not permit the animal to commit a nuisance on a sidewalk of any public place." Heavy fines could be levied on miscreant owners who failed to clean up the residue of their dogs' digestive achievements. Few tickets were actually issued by busy police, who had other tasks to perform more urgent than enforcing this virtually unenforceable law. But this did not prevent the ordinance from becoming an extremely effective tool in legitimizing and mobilizing intense social pressure against scofflaws. Middle-of-the-night outliers might occasionally escape public shaming, but only the bravest dared permit their pets to foul by day. As a New Yorker without a dog and with egregiously careless walking habits, I personally welcomed and very much benefited from the sudden cleansing of streets and sidewalks. The dog laws spread quickly across the country and world. Outrage at one very big Great Dane in one very small town had sparked a near-universal reversal in millennia-old dog-owning habits. Seemingly impregnable now, our much more important societal delusions may become equally pregnable in the future, once we cross a tipping point of public awareness and outrage.[1]

Litterbugs

Many of you are too young to know how disgustingly ubiquitous litter can be. When I was a kid, the world was a pigsty, with litter strewn everywhere. Movie theaters, parks, and beaches resembled garbage dumps; city streets and gutters were clogged with refuse; cigarette butts, lit and unlit, were thoughtlessly cast to the winds; wads of gum ambushed the unwary shoe; and people even threw stuff out of cars on highways, without giving it a second thought. The average American produces four pounds of trash a day— enough to create quite a mess if spread around liberally. Littering seemed to be an inviolable human right, inherent and inevitable in modern life, and just as unlikely as our current societal delusions to be addressed as a soluble problem.

The Keep America Beautiful campaign changed all that—and quickly. A community improvement organization with a small budget but big ambitions, it managed within just one decade to create a sea change in public attitudes toward, and behavior with, litter. There were three keys to its success. Most important was a just cause whose time had come. Second was an army of true-believing grassroots volunteers, organized into more than one thousand local chapters across the country. And most effective was its brilliantly shaming advertising campaign—a masterpiece of public education, propaganda, and prodding persuasion. *Litterbug* entered the language as a neologism of opprobrium and disgust. "Every Litter Bit Hurts" condemned even the smallest carelessness or disrespect.

And the one-minute-long "Crying Indian" television commercial changed the world. It is definitely worth googling and still chokes me up forty-five years after I first saw it. A solemn

Native American is paddling down a junk-filled stream whose banks are lined with polluting factories. He comes ashore, walks through more piles of junk, and approaches a highway. A passing motorist carelessly throws out yet more junk, which lands un-ceremoniously at his feet. A single, quiet tear forms on his stoical face as the narrator says, "People start pollution, people can stop it." The TV spot was a sensation, launching what has become an annual worldwide Earth Day. An "ecology flag," patterned on the American flag, but with green and white stripes, made littering unpatriotic. Follow-up campaigns (the Clean Community System and the Great American Cleanup) have kept the momentum going, cleaning up America one small litter piece at a time.

There is another, less idealistically uplifting (but practically crucial) moral twist to this complicated story. Keep America Beautiful is now, and always has been, funded in part by the food, tobacco, and beverage corporations—the despicable producers of much of the disposable stuff that winds up as litter. Solving the litter problem required an alliance both holy and unholy. Holy because it was far better to keep the crap off the streets. Unholy because it would be smarter and cheaper to make less crap. As evidence of the dark side, Keep America Beautiful opposes bottle deposits (even though they promote recycling and sustainability) and supports burning trash (even though it fouls the environment). A cynic might view Keep America Beautiful as no more than a deceitful corporate greenwashing device, designed to switch re-sponsibility for keeping a clean environment away from industry and on to consumers. It would be much better to have policies that encourage corporations to produce less trash, especially since the trash they produce inevitably winds up polluting the air or our oceans and rivers, or as landfill. In effect, Keep America Beautiful encourages good individual citizenship that succeeds in giving us clean streets, but it also enables corporations to be bad, dirty, and wasteful citizens. The larger moral: it is impractical to think we can change the world without corporate buy-in, but naïve to depend on corporate good intentions. The goal must be to create win-win situations where incentives are aligned. We must maintain the momentum on individual responsibility for eliminating litter, but also tax businesses for producing it.[2] This approach translates well to current efforts aimed at reducing the carbon-emission-caused global warming. Using fuel much more efficiently indirectly saves our environment by directly saving money for the corporations. For now, under Trump's polluting regime, we must rely on corporations to fight global warming and they will do so only if there is a profit in it.

Bottle-Fed

Until the early 1940s, when I was born, almost all mothers breast-fed their babies—just as mammals had been doing quite naturally for tens of millions of years. Cow milk substitutes had been around for a hundred years for mothers without milk or babies who wouldn't suckle, but their use risked causing vitamin deficiency or infection. Three interacting societal changes resulted in a sudden switch from breast to bottle. First, the war sent mothers to work, making breastfeeding inconvenient. Second, better formulas and safer delivery systems were developed. Third, and most important, formula makers such as Nestlé initiated a massive misinformation campaign persuading mothers that their artificial products produced healthier babies than did mother's milk. Also lurking in the background was that mothers were encouraged to believe that protracted sucking might have a detrimental effect on the beauty of their breasts. By 1970, only a fourth of mothers be-gan breastfeeding their babies at birth and almost none continued beyond two months. Nestlé had defeated nature.

But this victory of advertising artifice over common sense was short term. Four forces pushed back and restored breastfeeding to its rightful place in the natural scheme of things. First, scientists discovered many invaluable and irreplaceable ingredients in mother's milk that help babies resist infection and avoid allergies. Second, studies showed that breastfeeding helps forge a stronger and more secure mother–baby bond. The facts had made crystal clear that nature was indeed better for babies than Nestlé. Third, the formula makers overreached and were caught red-handed—outrageously propagandizing against mother's milk in third-world countries, even after the science had clearly contradicted their trumped-up case. The well-publicized exposés also opened the eyes of mothers in the developed world, who came to realize that they had previously been similarly snookered. And finally, and most important, La Leche (and similar organizations) worked hard to reeducate mothers about the health advantages of breastfeeding and to remind them of its inherent beauty and deep satisfaction as part of the human experience. Breastfeeding has made a dramatic comeback since the 1970s. Now three-fourths of mothers breastfeed at birth, half at six months, and one-quarter at one year. Mothers, not the medical profession, led this change in attitude. Moral of story: You can fool some women all the time, almost all women some of the time, but not all women all the time. And when women have a choice between breast shape or baby health, most do what's right for the next generation.[3]

Buckle Up

Cars didn't even have seat belts when I first began driving in 1960. By 1968, laws were passed that required carmakers to include seat belts, but most of us still didn't bother using them. Soon, graphic television ads showing the catastrophic impact of car crashes began to appear, especially emphasizing the value of seat belts even on the short-haul, low-speed trips close to home, where most accidents occur. But I, like almost everyone, still paid no attention and chose to continue driving free of restraint. Then, in 1984, New York became the first state to make seat belt use compulsory, with heavy fines for offenders. Before long, every state except New Hampshire required that all passengers use seat belts. Rates of use that were only 10 percent before the mandatory laws were passed are currently almost 90 percent. It has now become almost second nature for everyone, except the most irresponsible, to buckle up before driving—a true sea change in public attitudes and behaviors. Few people got fines, but most got the point—that seat belts save lives, an estimated 13,000 lives each year (most dramatically by eliminating the risk of being ejected from the car, which is a death sentence 75 percent of the time). More than half the people in fatal accidents are unbelted and the risk of death is cut in half when belts are worn.[4]

Despite this compelling case for belting, there are still some die-hard seat belt deniers. Their claim—that seat belts encourage more dangerous driving—has repeatedly been proven false. And libertarian arguments against mandatory use ignore the fact that society usually winds up paying most of the considerable costs of medical care and disability. The trajectory of seat belt use has been rapid—from not existing at all; to existing, but not being used; to becoming an essential and automatic part of American life—all in less than fifty years. This proves that sensible laws and regulations can overcome indifference, can change behavior, and can save lives. And that civil libertarian concerns should always be balanced against public health risks. We will someday be applying the very same logic to commonsense gun control— but only after politicians free themselves from unholy bondage to the National Rifle Association and begin to privilege public safety over gunmaker profit. The parallel between seat belts and gun control should be obvious to anyone not blinded by radical NRA rhetoric about an unlimited constitutional freedom to bear guns. If, to save lives, we place reasonable restrictions on driving, why should we do anything less for equally dangerous gun ownership?

Big Tobacco

I worked for eight years at Duke University—a great school made great by tobacco money. Now Duke prohibits smoking virtually everywhere on campus. Thirty years ago, Big Tobacco controlled Washington and its politicians. Now it is a toothless tiger and political pariah. Big Tobacco once ruled the media with seductive advertising—virile men and sexy women wielding cigarettes like sex organs. But then smoking ads were banned and antismoking ads flourished, featuring withered, wrinkled men and women struggling for every breath, hacking up sputum, clearly at death's door, disgusting to look at. Now there are few smoking ads at all, pro or con, because smoking is almost irrelevant. When I was a teenager, the height of cool was learning how to inhale. Now smoking is seen as a dirty and disgusting habit. When I was a young doctor, most doctors smoked. Now none do. Smoking used to be a norm. People smoked in bars and restaurants, on planes and trains, in theaters and movies, on hospital wards, in hotel rooms, in public restrooms, almost everywhere. I sometimes even caught hospital patients smoking while attached to their oxygen machines. Now it is becoming increasingly impossible to find a venue where it is legal to smoke, and completely impossible to find one where it is cool. Smoking is prohibited in almost all public places, indoors and out, even on the beach or in parks.

The Duke family pioneered, and for a long time dominated, the cigarette business. In the years after the Civil War, they were obscure Durham tobacco planters and traders. In the late 1800s they obtained an exclusive license on the world's first automatic cigarette-making machine. Within just a few decades, they parlayed their worldwide monopoly of machine-made cigarettes into a fabulous fortune. At its peak, almost half of all adults and two-thirds of men smoked and cigarette smoking had become one of the great public health disasters of all time—responsible for almost five hundred thousand deaths a year in the United States and five million worldwide. The first epidemiological studies connecting cigarettes and lung cancer were done in Germany in the 1920s. But, lost in the shuffle during the Nazi period, they had no impact. In the 1950s, Richard Doll independently and conclusively nailed the connection between tobacco and death. The British government took heed and issued the first public warning.

In the United States, there ensued a "Thirty Years' War" between the increasingly alarmed reports compiled by the surgeon general and the stubborn stonewalling of Big Tobacco. Advancing science was making it increasingly indisputable that smoking caused not just lung cancer, but also many other forms of cancer; increased risk for most cardiovascular diseases; and was implicated in a whole array of other illnesses. The tobacco industry responded by setting up its own fake "Research Council," intended to distract a gullible public with phony findings meant to discredit the work of the real scientists. Exactly the same game plan is followed today with stunning cynicism by energy companies and their toady politicians.

The tipping point for tobacco occurred in the late 1980s and early 1990s, when the surgeon general issued a particularly scathing report about nicotine addiction and its deadly consequences. The industry executives made fools of themselves on television at congressional hearings, mobilizing public outrage. And a class-action lawsuit confirmed, with internal industry documents, that Big Tobacco had been deliberately misleading the public about the dangers of smoking. The enormous fines it produced provided billions of dollars that funded effective antismoking campaigns. Then the clincher was compelling evidence that even passive inhalation of someone else's cigarette smoke could lead to deathly disease and that children were particularly vulnerable. Smokers were not just pathetically self-destructive, they were now seen as aggressively toxic to others. The sudden decline of cigarette smoking to 17 percent of the population is the greatest health triumph of my lifetime, far surpassing in saved lives all the combined efforts of high-tech medical treatments.[5] The fall of once-mighty Big Tobacco, defeated by a small band of antismoking scientists and advocates, should be a model and morale builder for the small and embattled band now challenging mighty Big Energy (and corrupt politicians) in the crucial fight for a livable planet.

The Ozone Hole

The ozone layer is a glorious climate change success story—and another perfect model for curing our other more severe pollution problems. Forty years ago, scientists discovered that aerosol sprays were releasing large quantities of chlorofluorocarbons (CFCs) into the stratosphere. This was a frightening health risk because CFCs degrade the ozone layer, which is needed to block ultraviolet radiation from reaching ground and people level. Without sufficient ozone in the atmosphere, rates of cancer and cataracts would skyrocket. Scared consumers responded well by boycotting CFC-containing products. The big chemical companies acted badly in the usual ways—stonewalling, lobbying, denying the science, smearing the scientists. Spurred by public concern, Congress was then able to pass protections to restore the ozone layer—an almost unimaginable feat in our day of complete congressional gridlock and unprecedented subservience of Congress to corporate, not public, interests. Also incredible, given our country's subsequent loss of credibility and moral standing, our State Department was able to successfully convince many other countries to follow suit with stiff laws to protect the ozone layer.

Things looked very promising until the Reagan administration came to Washington. In an ironic Newspeak twist, his appointees to the Environmental Protection Agency (like Trump's) seemed determined to do everything in their power to destroy the environment they were responsible for protecting—they never met a chemical pollutant or polluter they didn't like. But Reagan's reflexive climate denial could not withstand the indisputable discovery of a growing hole in the Antarctic ozone layer—a tangible and visible threat that roused the entire world to action. The 1989 Montreal Protocol banned all use of CFCs and has saved our ozone layer.[6]

Could we beat global warming if we mounted a similar comprehensive international response to CO_2 pollution? Sure, but it is a much steeper hill to climb. Curing the ozone hole was relatively cheap and easy. Curing CO_2 is remarkably expensive and would require great sacrifices. And we have nothing comparable to the dramatic and palpable pictures of the Antarctic ozone hole to convince dead-end climate-change doubters that global warming is no hoax. But these are differences only in quantity, not in kind. The world will probably act sensibly when the CO_2 greenhouse effect begins causing regular, destructive, expensive, and predictable catastrophes. The unanswerable question always is: Will it be too little, too late? Too bad there is no visual equivalent of the ozone hole to get us started now, while there is still time. The graphs charting carbon dioxide concentrations over time are pretty scary but don't carry nearly the same emotional wallop.

Acid Rain

Acid rain was first described as a curiosity almost five hundred years ago, but practical interest perked only when the smokestacks of the Industrial Revolution started belching out fumes containing sulfur dioxide. Joined with atmospheric water vapor, the resulting sulfuric acid falls downwind as acid rain. Dan Smiley compiled the most complete record of its fluctuations during most of the twentieth century. A dedicated amateur naturalist working in the tradition of Henry David Thoreau, Smiley sought to measure every measurable fact about the land surrounding his beloved resort in New York State—and he did so most days between 1931 until his death in 1989. Among Smiley's many measures was his reading of the daily acidity of the waters of Lake Mohonk, allowing him to prove conclusively that its pH was slowly but steadily dropping. His daughter, my friend Anna Smiley, sometimes tramped along with him on his appointed rounds.

Smiley's findings were confirmed all around the world, particularly in places near power plants burning sulfur-containing coal. Building taller smokestacks to protect the local neighborhood just spread the problem farther afield. In some heavily industrialized areas, pH dropped below 3, enough to kill plant and animal life, to peel the paint off bridges, and to damage the stones of historical monuments. The public became aware of acid rain in the 1970s through extensive media coverage and a devastating report from the National Academy of Sciences. Congress responded by strengthening the Clean Air Act and established an emissions cap-and-trade system that has reduced acid rain by 65 percent. The cost to business and consumers is just a bit more than $1 billion a year, less than one-fourth of the estimates made during the industry's fearmongering campaign against regulation. The European Union has taken even more aggressive action and has recorded even larger reductions. This is the good news.[7]

Now for the bad news—China is burning coal like there will be no tomorrow. Every five years, it adds new plants equal in capacity to all the current coal-firing plants in the entire United States. On average, one new plant has been opened every week for decades, each with life expectancy of forty years.[8] And other developing countries, playing the game of power generation catch-up, are also heavily dependent on cheap and dirty coal. Acid rain is the canary in the coal mine for the entire problem of carbon pollution on our planet.

Developing countries argue that because richer countries have caused most of the existing world pollution, they have already gained the wealth and should therefore bear the costs of preventing further pollution—even when it is generated outside their borders in the developing world. Developed countries naturally prefer a more blank-slate approach that avoids their having to transfer tens (perhaps hundreds) of billions of dollars to countries that have now become very tough economic competitors. The United States argues that China should use its accumulating great wealth to clean up its own mess. Meanwhile, the world is burning coal, while the bureaucrats fiddle. Negotiations have been frustrating, and meaningful action keeps getting kicked down the road. Compliance is hard to monitor and enforce. Sooner or later, there will be agreement—but again, it may come too late.

Civil Rights

When I was born in 1942, women had been voting for just twenty-two years; racial segregation was practiced rigidly in many states and in the U.S. Army; miscegenation and homosexuality were crimes in some jurisdictions, shames in all. Today the laws take it as given that women are equal to men;[9] that blacks are equal to whites;[10] that sexual and gender orientation are irrelevant to your rights as a citizen.[11] But just yesterday these were fighting words.

The Bill of Rights and Constitution had grandly alluded to "human rights," but only in the most general way, not in concrete terms that could protect the "civil rights" of any individual person. All men were certainly not equal before the law and blacks and women were not even recognized as citizens. Progress in ensuring legal equity has been slow, unstable, shamefully incomplete, often unfair, and almost always tumultuous—but in the past sixty years, also amazingly successful.

Only under the shadow of a hideous Civil War did Congress pass the Fourteenth Amendment of the Constitution, which stated: "All persons born or naturalized in the United States, and subject to the jurisdiction thereof, are citizens of the United States and of the State wherein they reside. No State shall make or enforce any law which shall abridge the privileges or immunities of citizens of the United States; nor shall any State deprive any person of life, liberty, or property, without due process of law; nor deny to any person within its jurisdiction the equal protection of the laws." The federal government was given powers, not previously within its purchase, to prevent state governments from discriminating, particularly against the newly freed slaves.

Nice on paper, but not at all effective in preventing the Jim Crow laws that succeeded in nullifying the Reconstruction. The battle for legal recognition of civil rights, badly lost in the nineteenth century, was hard and mostly successfully fought in the twentieth and twenty-first. Women, racial minorities, and LGBT people have obtained legal standing that would not have been imaginable in the relatively recent past. The results are, far too often, more de jure than de facto, but nonetheless, the forward march is not to be denied. Having a black president shows how fast and far we have come. Much remains to be done, but much has been achieved.

Tactics in pushing for civil rights have varied with time, issues, politics, economics, demographic circumstances, and the personalities of the players. But the strategy is always similar—to focus media, public, and legal attention on the injustice of the discrimination; on the basic humanity of those being discriminated against; and on the basic inhumanity of those doing the discriminating. When it comes to ensuring justice and tolerance, politicians and judges much more often follow public opinion than lead it. The most frequently used techniques to promote social change have been demonstrations, civil disobedience, grassroots organizing, voter drives, political pressure, legal pressure, and economic pressure through boycotts. The goal is to mobilize first a small group of the most committed individuals and, through them, the general public. Seeing the injustice on television helps (almost) everyone identify with, and feel empathy for, the victim.

Words and deeds have been synergistic, equally important, and interdependent in forcing changed perceptions and policies.

One tipping point toward the passage of the Civil Rights Act was Martin Luther King's galvanizing appeal to end racism—delivered on the steps of Lincoln's memorial, he inspired the two hundred fifty thousand people in attendance, and tens of millions more watching on television, with the words "I have a dream that one day this nation will rise up and live out the true meaning of its creed: 'We hold these truths to be self-evident, that all men are created equal.'" King had succeeded in shaming the United States into living up to its own self-proclaimed values. The other tipping point occurred before the march on Selma. Lyndon Johnson advised King that the propaganda value of even one egregiously outrageous act of discrimination would be needed to garner public sympathy and congressional support for the impending voting rights bill: "Get it on radio, get it on television, get it in the pulpits, get it in the meetings, get it every place you can; pretty soon, the fellow that didn't do anything but drive a tractor will say: 'That's not right, that's not fair.' And then, that'll help us in what we're going to shove through in the end." Johnson understood how people led policy.

Public opinion, though often appearing fixed, is in fact fickle and malleable—for the good or for the bad. The public is currently being manipulated by Trump and the big-money boys who prey on fear to allow the pursuit of their own narrow, selfish interests. Sadly, we lack a Martin Luther King to embody the contrary inclusionary message that our world is shrinking and that we are all in this together. Progress won't be easy, inevitable, or without setbacks—we take a big step forward when we elect a black president, then suddenly a big step back electing a racist one. The Trumps of the world won't suddenly disappear or become enlightened. Women, racial minorities, and the LGBT community have had to fight a continuing, long, and hard battle to change attitudes among those capable of attitude change. We must now fight equally long and hard to keep our planet habitable by establishing a global ethic of inclusiveness that embraces everybody alive today and protects the rights of those yet unborn.

We Are One Family

In most parts of the world, nation-states are a relatively new, and still extremely fragile, form of governance. Allegiance has generally been much more local. Hunters and gatherers felt loyalty to their small wandering groups. Larger political institutions became possible only after the agricultural revolution, when the accumulation of wealth allowed the accumulation of land and power. In most times and in most places, the unit of individual allegiance remained the extended family, the village, the tribe, and the religious group—not a nation.

Current nation-states vary in age from born yesterday to very young. The newest are only decades old, arising from the sorting out that occurred after the collapse of the Soviet Union and Yugoslavia; African countries are mostly fifty years old; "India" and Pakistan are only seventy years old; "Ireland" is fewer than one hundred; "Germany" and "Italy" fewer than 150; England, France, and Spain are barely five hundred; and even the oldest, China and Japan, have often in their histories been divided into warring states. Most of the new "nations" are awkward postcolonial creations with artificial boundaries drawn by colonial administrators for their own convenience, often with dubious logic and uncertain stability. Borders between "nations" routinely lump together many heterogeneous tribal and religious groups that don't want to be together, while dividing cohesive groups across artificial boundaries. The concept "Iraq" or "Syria" or "Somalia" or "Afghanistan" or "Sri Lanka" has more meaning to distant politicians than to the people on the ground.

Most of us take for granted that love of country is a natural and noble feeling. In fact, it is a relatively recent development in human history—not particularly natural, and often not particularly noble. The word *patriotism* is only three centuries old, introduced as part of the Enlightenment effort to replace religious institutions with secular ones. A romantic attachment to the state was meant to substitute for, and improve upon, loyalty to the church. Almost immediately there was backlash, most vividly expressed in 1775 by Samuel Johnson's extreme statement: "Patriotism is the last refuge of the scoundrel." Patriotism, like religion, has its valuable uses, but also has its dangerous misuses.

The powerful patriotism felt toward the United States is only about half as old as the country itself, an unexpected consequence of the brutal civil war that almost dissolved it. The United States had become one nation just four-score-and-five years before the secession that threatened to make it two, or many. It had been formed by the joining of thirteen formerly separate British colonies, widely divergent in their histories, demographics, economic systems, trading partners, laws, and traditions. The different colonies had found common cause in fighting a common enemy, but aside from language had precious little else in common. After struggling long and hard to free themselves from perceived tyranny by king and parliament, the new state governments were naturally fearful of establishing a strong homegrown central government. Their first postwar contract, the Articles of Confederation, was carefully written to preserve the maximum possible independence of each former colony, linking them with only the weakest bonds. The United States were thus first united in name only and proved ungovernable. *The Federalist Papers,* arguing for a more perfect union, inspired the Constitutional Convention that went far toward creating it. But the union was still far from perfect—the states still retained considerable freedom from president, Congress, and federal courts. Most citizens felt their greatest loyalty to their state and local institutions, not to the more abstract concept of United States.

The lack of clarity between states' rights and federal control made the Civil War inevitable. And only in the crucible of that war did the states truly become united. Before the war, the country was frequently referred to as "these United States" and the plural was almost always used to describe them (as in "these United States are"); after the war this usage was gradually replaced by the singular ("the United States is"); a small change in article, but a big change in meaning. Allegiance, even in the former Confederacy, gradually shifted from the state to the nation. Near-universal American patriotism was cemented by the Spanish-American War and World War I. Having a common enemy brought us closer together.

In the wake of the Nazi experience, Albert Einstein stated that "[n]ationalism is an infantile disease. . . . It is the measles of mankind." Being strongly nationalistic is now increasingly anachronistic and counterproductive—especially if loving my country means hating and fearing yours. Globalized problems require globalized solutions. Lest we sink our planetary ship while fighting over it, we must expand local loyalties so that they include the entire species.

The past century has launched three grand experiments in international cooperation: the League of Nations, which failed miserably; the United Nations, which is failing; and the much more promising European Union, which outgrew itself too quickly and now risks falling apart. This discouraging track record undercuts any possible utopian delusion that our species can easily cooperate to form "Team Earth" and save the sinking ship. Except in the most extreme of circumstances, pressing personal, corporate, and national self-interest will almost always trump what is in the best interest of all mankind. But we are now approaching those most extreme of circumstances. I am not naïve enough to believe we can ever become postnational out of love, but we might conceivably achieve much closer international cooperation out of fear. Local differences dissolve, and unexpected alliances form, when diverse interests are confronted by a threatening external enemy.

It is admittedly hard to be optimistic about world unity at a moment when Trump and Putin are doing their very best to create a maximum of world disunity and the European community shows signs of no longer being a community. But we must hope that the logic of necessity will prevail over the illogic of this moment. We can't possibly solve global problems without global solutions. Global warming is a global disaster that can only be avoided with global cooperation. Civil wars start within one country, but immigrants escaping from them are a destabilizing force everywhere. Epidemics know no national borders. And a failure of the banking system in one country can trigger an economic disaster in all countries. We already have numerous examples of successful international cooperation in politics, trade, banking, health, migration, arbitration, and international law. We also have numerous governmental and nongovernmental institutions that provide a ready infrastructure for global cooperation. The Internet ties our world in a tight communication embrace never before possible. English has become a lingua franca. People everywhere listen to the same music, see the same movies, wear the same clothes, and use the same smartphones. Never has it been so easy to conceive a Team Earth, once we can overcome local prejudices. We are one species in the process of destroying itself. It is only as one species that we can save ourselves. Fear of the consequences of a fractious future will be the best glue to cement a more cooperative present.

The faster our world recognizes that we are entering a crisis of existential proportions, the less likely it will overwhelm us. I believe that much greater world unity of purpose and action will likely soon become possible, indeed probable, when the peoples and leaders of the world become scared enough about the future to stop frenetically pursuing their selfish and short-term interests in the present.[12]

Geniuses and Jackasses

Yogi Berra, that great American folk philosopher, once quipped: "It ain't over till it's over." And in the big game of species survival, it still ain't over for the human race. Sure, we are living through a Trumpian dark age of breathtaking arrogance and ignorance (a simply awful combination); of heated dissension within the country and dangerous confrontations without; of science denial and special interest pandering; of policies and people promoting the dumbest and most dangerous version of every single societal delusion. We are behind in the late innings, but still stand a fighting chance to make things right. Once we get rid of Trump and the retrogressive Republican domination of government, new leadership will hopefully help us emerge from our current dystopia by facing reality and joining us together as a team.

My father's summary judgment of me, made shortly before his death ("mostly jackass, with a small spark of genius"), applies equally well as a more general statement about humanity. We are the best and worst of species—often jackasses who create terrible problems, but occasionally geniuses who create beautiful solutions. Evolution somehow saw fit to produce a Trump, but it also created Einstein and Shakespeare and Eleanor Roosevelt. Humanity has done, and continues to do, some amazingly stupid and self-destructive things. We are cursed with the original sins of greed, selfishness, and shortsightedness. But we are also extremely resourceful and resilient creatures. Our better angels do sometimes win out against even very long odds, especially when the chips are down. Right sometimes does triumph over entrenched might.

And our genetic endowment is a mixed bag, with lots of good stuff along with the bad. We are pack animals, preprogrammed to feel deep satisfaction in furthering the survival of our pack. We need to be a part of something bigger, responsible not only to ourselves, but also to family, community, country, and now (I hope) world. When threatened, pack animals form a circle of protection to defend themselves and take care of their own. If we are to survive as a species, we must widen our conception of who is in our pack and who is the enemy outside. Our circle of protection must expand to embrace the entire circumference of the earth, all its places, all its diverse people, all its many species. "Foreigners" were once our rivals; now they must be seen as siblings and cousins: we cannot hurt them without hurting ourselves; helping them helps us and adds meaning to our lives.

I am not a naïve believer that universal brotherhood will come easily. Anyone who knows anything about human history realizes it has been a sorry pageant of constant, petty, destructive warfare. But there is also comfort in knowing that the step toward becoming good earth citizens is a natural one, requiring no more than extending the definition of our pack. And as Aristotle put it: "A common danger unites even the bitterest enemies." We certainly face a common set of self-inflicted dangers and must be united if we are to overcome the existential threats we have created. We can control our fate only if we can control ourselves.

Can We Learn from History?

"History is a nightmare from which I am trying to awake." James Joyce's despair is understandable, but must be avoided. History is certainly littered with civilizations that failed to wake up to what should have been evident warnings—blithely assuming that nature would be endlessly bountiful, that war would end in glorious success, and that more was always better. Why imagine that we are any different? The smart money might bet that the one thing we learn from history is that we don't learn from history—and that our species, having shot so high, will have a most catastrophic, Humpty-Dumpty fall.

My head tells me that this is a selfish, blind, and stupid world unable to act responsibly on its all-too-obvious problems—massive overpopulation, environmental degradation, resource depletion, inequality, climate change, and all the rest. The appetites that served us well in the million-year march to this moment were adaptive when we were few and the world was large, but are deadly now that we are many and the world is small. My head is a realist fearing the worst.

My heart prefers Santayana's more optimistic position that those who don't know the past are doomed to repeat it—hopefully implying that those who do know the past can perhaps avoid its pitfalls. Fatalism is a dangerously self-fulfilling prophecy—if we feel trapped by our past, we will be passive in rewriting our future. And our past is also filled with useful messages and hopeful models, not just depressing reminders of blown opportunities. We can learn from our previous triumphs, not just blindly repeat our past mistakes.

Mark Twain had it right when he said that history doesn't repeat itself, but it sure does rhyme. The contingencies are always too complicated to predict exactly how things will turn out. Historical forces created a Trump, but his winning or losing an incredibly consequential election was very much a function of fickle chance (assisted by the meddling of Vladimir Putin). Things even out in the long run, but there isn't always a long run. I trust in humanity enough to believe it inevitable that we will finally wise up to our problems and create clever political and technological solutions for them. I have much less trust that we will awake in time from the nightmare of our history. The world's problems escalate in difficulty the longer we neglect them. I fear the worst, but hope for the best.

Do You Believe in Miracles?

It is perhaps my personal delusion to think we will face reality, overcome our societal delusions, and save the day. I like long shots and we have no choice but to bet on this one. And sometimes I believe in miracles. In the 1980 Winter Olympics in Lake Placid, New York, the U.S. men's hockey team was coached by Herb Brooks, a stern taskmaster unpopular with his players, mostly college kids in their early twenties. The U.S. team was good enough to make it to the Olympic semifinals, but was given absolutely no chance against their next opponent, the perennial world champion Russian team, consisting of some of the world's best players, who had been together for years and were in the prime of their hockey lives. Just two weeks before, the Russians had overwhelmed the Americans 10–3 in an exhibition match that was even more lop-sided than the final score. No one except Herb Brooks believed in his untried and much less talented U.S. team. In the locker room before the game, he told his kids: "Great moments . . . are born from great opportunity. And that's what you have here, tonight, boys. That's what you've earned here tonight. One game. If we played 'em ten times, they might win nine. But not this game. Not tonight. Tonight, we skate with them. Tonight, we stay with them. And we shut them down because we can! Tonight, *we* are the greatest hockey team in the world." And they were, achieving a ridiculously implausible comeback win. As the clock ran out, the announcer Al Michaels shouted passionately: "Do you believe in miracles?" Usually I don't, but sometimes they happen. We all need to believe in the miracle of Team Earth.

Epilogue

Whither Mankind?

By strange coincidence, I began revising this book while on a visit to Easter Island (Rapa Nui)—a place that tells us all we need to know about human greatness and human fallibility. The remarkable civilization created on this tiny island displayed brilliant human inventiveness but was quickly destroyed by dumb human blindness. Pride preceded and, partly, caused the catastrophic fall. The few dozen Polynesians who colonized Easter Island eight hundred years ago must have been among the smartest people who ever lived. They successfully navigated more than two thousand miles of open ocean (against prevailing currents) to colonize one of the most isolated and unpromising places on earth. The Rapa Nui were among the few peoples in the history of the world ever to independently develop a writing system. And, uniquely, they achieved this intellectual feat in complete isolation—without the usually necessary trade in ideas that comes with the trade of goods. Their tiny, not very fertile island was converted into a highly efficient garden spot, seducing them into an exponential population spurt from a few dozen people to more than ten thousand.

The agricultural abundance led, as it always does, to a highly stratified, overpopulated society, preoccupied with monument building to establish the power and prestige of those presiding at the top. Never before has such a small place created so many big monuments—one thousand statues that are a wonder of engineering, artistic creativity, and craftsmanship, but also of human greed and vainglory. In the early years, they were just a few feet tall, weighed just a few tons, and were roughly hewn. Size progressively increased over time, with the last monuments being the grandest in ambition, more than forty feet tall and weighing eighty tons. Each statue had to be walked many miles from quarry to final resting place, using an ingenious towing system that maximally leveraged human pulling power. Tellingly, about half of them, the biggest, were still being carved even as the civilization was collapsing. They stand now as a haunting reminder of human greatness, but also of its lack of wisdom and self-restraint. Easter Island is now one of the saddest places on earth—barren scrub, dotted with the most magnificent reminders of lost past glory.

The tragedy of Easter Island is shared by many past civilizations, in all epochs, in all parts of the world. The pattern is depressingly familiar—a rise to great heights followed by collapse, always caused by overpopulation, resource depletion, and climate change. Historians and journalists generally ignore the underlying causes and focus only on the surface results (the civil wars, the invasions, the political intrigues). These make for more dramatic narrative but miss the real story that demography and climate are destiny. Every great civilization has run the same cycle—ultimately doomed by its own success in breeding and technology. We push our limits until our environment is degraded, our natural resources are dwindled, and the rains stop falling. If our civilization doesn't soon come to its senses, archaeologists of the future will wonder how we could have been simultaneously both so smart and so dumb. Easter Island is the perfect exemplar and metaphor for man's seemingly inescapable fate. I loved being there, but sometimes a metaphor can make you cry. We learn from Easter Island or we repeat Easter Island.

There are two possible scenarios for our next half century— our species will either come together or tear itself apart. Our biological makeup and social structures are completely compatible with either result. We have "bad angel" genes and institutions that propel us to greed, competition, aggression, and shortsighted decision making. But we also have "better angel" genes and institutions that promote altruism, sharing, responsibility, and rational decision making. The basics of our human nature are stable, but the way we behave is very situation-specific. Yesterday's berserking Vikings are today's Nordic rationalists.

Worst case: Water and energy resources are depleted. Overpopulation and climate change make parts of the world unlivable. Tribal entities arm themselves to the teeth and fight to the death in near-perpetual wars to steal or defend bigger slices of an ever-shrinking pie. There is spiraling anarchy, chaos, famine, disease, and an inability to respond to increasingly frequent natural and man-made disasters. This scenario has been repeated over and over again as the endgame for collapsing great civilizations. And we have had, and are having, many recent and current foretastes of this possible bleak future.

Best case: The big economic powers—the United States, China, and the European Union—realize that they are in the same sinking boat and must cooperate closely and fully on the major problems facing humanity. They drastically shrink military budgets to fund the infrastructure programs intended to reduce waste and pollution and to restrain population growth. Together the Big Three provide adult supervision and a model to the rest of the world—pulling along cooperating countries with financial incentives and technology transfers and pushing away noncomplying countries with tariffs, sanctions, and boycotts. They, along with the less wealthy and developed countries, join a worldwide parade toward a sustainable planet. This is not as pie-in-the-sky as it may sound. There is every reason to expect this will happen, because, in part, it already has happened—just too little, but hopefully not too late.

We will either join together to help solve the world's problems or we will worsen them with the war of all against all. The future is being written now and the current chapter could not possibly seem more bleak—the United States deeply immersed in its most dystopian dark age. Trump may well be a tipping point—presaging either future worldwide democratic decay and environmental catastrophe, or the crisis before the fever of societal delusion finally breaks. The contingencies of history are often precariously balanced and very evenly matched, the results of the endgame uncertain and unpredictable. Trump has taken away any middle ground—you are either for him and support societal delusion or you are against both. This is no time for responsible people to hide behind passivity or fatalism.

I am fully aware that no single book can do much to save our brave new world, but we all have our small roles to play in achieving a saner society. As Edmund Burke put it: "Nobody made a greater mistake than he who did nothing because he could do only a little." I think we must approach the future with open eyes, open minds, and open hearts—fearing its worst, but working with all our might to bring out its best. We owe no less to our children and grandchildren and their children unto the generations. Imagine the day when your grandchild may ask: "What were you doing when Trump destroyed our world?"

Acknowledgments

My wife, Donna Manning, was a full partner in every word—and provided all of this book's heart and most of its brains. My agent, Carrie Kania, gave me the idea and the inspiration to risk tilting at windmills. My editor, Peter Hubbard, pointed the way, made wonderful suggestions, and kept me on the path. My children, Craig and Bob; my grandchildren, Tyler, Olivia, Angelina, Jared, and Jack; and their unborn children provided my motivation. My father, Joe Frances, was my model.

Notes

Chapter 1: Confronting the Facts of Life

1. James Hansen, *Storms of My Grandchildren* (New York: Bloomsbury, 2009).

2. Neela Banerjee, Lisa Song, and David Hasemyer, "Exxon's Own Research Confirmed Fossil Fuels' Role in Global Warming Decades Ago," *Inside Climate News,* September 16, 2015, https://insideclimatenews.org/news/15092015/Exxons-own -research-confirmed-fossil-fuels-role-in-global-warming.

3. Thomas Malthus, *An Essay on the Principle of Population, As It Affects the Future Improvement of Society* (London: 1798), http://www.esp.org/books/malthus/population /malthus.pdf.

4. Max Roser and Esteban Ortiz-Ospina, "World Population Growth," OurWorld InData.org, 2016, https://ourworldindata.org/world-population-growth/.

5. Timothy B. Gage and Sharon DeWitte, "What Do We Know About the Agricultural Demographic Transition?" *Current Anthropology* 50 (2009): 649–55.

6. "Syria Population," accessed December 27, 2017, http://countrymeters.info/en /Syria.

7. John Hudson, "UN Envoy Revises Syria Death Toll to 400,000," *Foreign Policy,* April 22, 2016, http://foreignpolicy.com/2016/04/22/u-n-envoy-revises-syria -death-toll-to-400000/.

8. "World Population vs. World Oil Production," YouTube video, 3:02, posted by "R. E. Heubel," June 9, 2011, https://www.youtube.com/watch?v=8vljIei1PwM.

9. R. W. Allmendinger, "Peak Oil?," Cornell University Energy Studies in the College of Engineering, 2007, http://www.geo.cornell.edu/eas/energy/the_challenges /peak_oil.html.

10. Brian Handwerk, "Underground 'Fossil Water' Running Out," *National Geographic News,* May 8, 2010, http://news.nationalgeographic.com/news/2010/05/100505 -fossil-water-radioactive-science-environment/.

11. *Water Encyclopedia,* "Ogallala Aquifer," 2016, http://www.waterencyclopedia.com /oc-po/ogallala-aquifer.html.

12. Frans de Waal, "Moral Behavior in Animals," TEDxPeachtree video, 16:52, filmed November 2011, http://www.ted.com/talks/frans_de_waal_do_animals_have_morals.

13. Ricardo Fuentes-Nieva and Nick Galasso, "Working for the Few," *Oxfam Briefing Paper—Summary,* January 20, 2014, https://www.oxfam.org/sites/www.oxfam .org/files/bp-working-for-few-political-capture-economic-inequality-200114 -summ-en.pdf.

14. Alyssa Davis and Lawrence Mishel, "CEO Pay Continues to Rise as Typical Workers Are Paid Less," Economic Policy Institute, Issue Brief #380, June 12, 2014, http://www.epi.org/publication/ceo-pay-continues-to-rise/.

15. "Power in Numbers: Lobbyists Have Congress Covered," Face the Facts USA, January 14, 2013, http://www.facethefactsusa.org/facts/power-numbers-lobbyists -have-congress-covered.

16. Sean Kane, "The Human Race Once Came Dangerously Close to Dying Out— Here's How It Changed Us," *Tech Insider,* March 18, 2016, http://www.business insider.com/genetic-bottleneck-almost-killed-humans-2016-3.

17. Michael White, "What Our Genes Tell Us About Race," *Science 2.0,* November 5, 2008, http://www.science20.com/adaptive_complexity/what_our_genes_tell_us _about_race.

18. Luigi Luca Cavalli-Sforza, Paolo Menozzi, and Alberto Piazza, *The History and Geography of Human Genes* (Princeton, NJ: Princeton University Press, 1994).

19. "How Many Species on Earth? About 8.7 Million, New Estimate Says," *Science Daily,* August 24, 2011, https://www.sciencedaily.com/releases/2011/08/1108231 80459.htm.

20. Rob Jordan, "Stanford Researcher Declares That the Sixth Mass Extinction Is Here," *Stanford News,* June 19, 2015, http://news.stanford.edu/2015/06/19/mass -extinction-ehrlich-061915/.

21. "Current Extinction Rate 10 Times Worse Than Previously Thought," *IFL-Science!* http://www.iflscience.com/plants-and-animals/current-extinction-rate-10 -times-worse-previously-thought/.

22. Douglas Main, "How Endangered Species May Fare Under Trump," *Newsweek,* January 12, 2017, https://www.newsweek.com/how-endangered-species-may-fare -under-trump-542019.

23. Karin Brulliard, "USDA Abruptly Purges Animal Welfare Information from Its Website," *Washington Post,* February 3, 2017, https://www.washingtonpost.com /news/animalia/wp/2017/02/03/the-usda-abruptly-removes-animal-welfare -information-from-its-website/?utm_term=.a6a477cbc148.

24. George Orwell, *1984* (London: Secker & Warburg, 1949).

25. Matt Shuham, "The NSA Leaks: A Summary," *Harvard Political Review,* August 11, 2013, http://harvardpolitics.com/united-states/the-nsa-leaks-a-summary/.

26. Dana Priest and William M. Arkin, "The Secrets Next Door," *Washington Post,* July 21, 2010, http://projects.washingtonpost.com/top-secret-america/articles/secrets -next-door/print/.

27. Edwin Black, *IBM and the Holocaust: The Strategic Alliance Between Nazi Germany and America's Most Powerful Corporation* (Lanham, MD: Crown Books, 2001).

28. Amanda Li, "9 Pros and Cons of Surveillance Cameras in Public Places," Reo-link, April 16, 2016, https://reolink.com/pros-cons-of-surveillance-cameras-in -public-places/.

29. Michael P. Lynch, *The Internet of Us* (New York: Norton, 2016).

30. "Gun Violence by the Numbers," Everytown, https://everytownresearch.org /gun-violence-by-the-numbers/.

31. Marty Langley and Josh Sugarmann, "Firearm Justifiable Homicides and Non-Fatal Self-Defense Gun Use: Analysis of Federal Bureau of Investigation and National Crime Victimization Survey Data," Issuelab, April 1, 2013, http://gun violence.issuelab.org/resource/firearm_justifiable_homicides_and_non_fatal_self

_defense_gun_use_an_analysis_of_federal_bureau_of_investigation_and_national _crime_victimization_survey_data.

32. "Defense and Arms—Statistics & Facts," Statista, https://www.statista.com/topics /1696/defense-and-arms/1.

33. Ray Kurzweil, *The Singularity Is Near: When Humans Transcend Biology* (New York: Penguin, 2006).

Chapter 2: Why We Make Such Bad Decisions

1. Plato and Stephen Scully, *Phaedrus* (Newburyport, MA: Focus/R. Pullins, 2003).

2. Charles Darwin, *The Expression of the Emotions in Man and Animals,* (London: John Murray, 1872), http://gruberpeplab.com/3131/Darwin_1872.pdf.

3. Sigmund Freud, *An Outline of Psycho-Analysis* (New York: Norton, 1949).

4. Michael Lewis, *The Undoing Project: A Friendship That Changed Our Minds* (New York: Norton, 2016).

5. Thucydides, *History of the Peloponnesian War,* trans. Rex Warner (Harmondsworth, England: Penguin Books, 1972).

6. Aristotle, *The Politics and the Constitution of Athens,* (Cambridge, UK: Cambridge University Press, 1996).

7. Edward Gibbon, *The History of the Decline and Fall of the Roman Empire* (New York: Harper & Brothers, 1836), https://www.gutenberg.org/files/25717/25717-h/25717 -h.htm.

8. Jared Diamond, *Guns, Germs, and Steel* (New York: Norton, 1997).

9. Niles Eldredge, "Darwin's *Other* Books: 'Red' and 'Transmutation' Notebooks, 'Sketch,' 'Essay,' and *Natural Selection,*" *PloS Biology* 3, no. 11 (November 2005): e382, https://www.ncbi.nlm.nih.gov/pmc/articles/PMC1283389/.

10. John Locke, *An Essay Concerning Human Understanding,* (London, 1690), https:// ebooks.adelaide.edu.au/l/locke/john/l81u/contents.html.

11. Daniel Kahneman, *Thinking, Fast and Slow* (New York: Farrar, Straus & Giroux, 2011).

12. Nick Lane, *Life Ascending: The Ten Great Inventions of Evolution* (New York: Norton, 2016).

13. Jean-François Gariépy, "Why Do Nervous Systems Use Slow Voltage Changes Rather Than Fast Electric Currents Along Wires?" BrainFacts.org, August 19, 2014, http://blog.brainfacts.org/2014/08/why-do-nervous-systems-use-slow-voltage -changes-rather-than-fast-electric-currents-along-wires/#.WGQBNcuIbqA.

14. Michael W. King, "Biochemistry of Neurotransmitters and Nerve Transmission," themedicalbiochemistrypage.org, last modified November 23, 2016, http:// themedicalbiochemistrypage.org/nerves.php.

15. Stephanie Pappas, "Oxytocin: Facts About the 'Cuddle Hormone,'" *Live Science,* June 4, 2015, http://amp.livescience.com/42198-what-is-oxytocin.html.

16. R. J. R. Blair, "Considering Anger from a Cognitive Neuroscience Perspective," *Wiley Interdisciplinary Reviews Cognitive Science* 3, no. 1 (January–February 2012): 65–74, http://doi.org/10.1002/wcs.154/.

17. Thomas Hobbes, *Leviathan* (New York: Penguin Classics, 1982).

18. Walter Mischel, Yuichi Shoda, and Monica Rodriguez, "Delay of Gratification in Children," *Science* 244, no. 4907 (1989): 933–38.

19. T. Sharot et al., "Neural Mechanisms Mediating Optimism Bias," *Nature* 450, no. 7166 (November 1, 2007): 102–5.

20. *Optimism: A Report from the Social Issues Research Centre,* February 2009, http://www.sirc.org/publik/Optimism.pdf.

21. David Hecht, "The Neural Basis of Optimism and Pessimism," *Experimental Neurobiology* 22, no. 3 (2013): 173–99.

22. Frieder R. Lang et al., "Forecasting Life Satisfaction Across Adulthood: Benefits of Seeing a Dark Future?" *Psychology and Aging* 28, no.1 (2013): 249–61.

23. Catherine A. Hartley and Elizabeth A. Phelps, "Anxiety and Decision-Making," *Biological Psychiatry* 72, no. 2 (2012), https://www.ncbi.nlm.nih.gov/pmc/articles/PMC3864559/.

24. D. R. Oxley et al., "Political Attitudes Vary with Physiological Traits," *Science* 321, no. 5896 (September 19, 2008): 1667–70, https://www.ncbi.nlm.nih.gov/pubmed/18801995.

25. Philip A. Gable, Bryan D. Poole, and Eddie Harmon-Jones, "Anger Perceptually and Conceptually Narrows Cognitive Scope," *Journal of Personality and Social Psychology* 109, no. 1 (2015): 163–74, https://pdfs.semanticscholar.org/eab3/04ad3c752988265f45d372f3e0e59324ef64.pdf.

26. Iddo Gal, "Adults' Statistical Literacy: Meanings, Components, Responsibilities," *International Statistical Review* 70, no. 1 (2002): 1–52, https://www.stat.auckland.ac.nz/~iase/cblumberg/gal.pdf.

27. Kendra Cherry, "What Is a Confirmation Bias? Examples and Observations," *Verywell,* June 22, 2016, https://www.verywell.com/what-is-a-confirmation-bias-2795024.

Chapter 3: American Exceptionalism

1. Ann Richards discussing George H. W. Bush in 1988.

2. Alexis de Tocqueville, *Democracy in America: The Complete and Unabridged Volumes I and II* (New York: Bantam Classics, 2000).

3. Thomas More, *Utopia* (New York: Penguin, 1984).

4. William Shakespeare, "The Booke of Sir Thomas Moore" (unpublished manuscript, c. 1601–04), British Library collection, https://www.bl.uk/collection-items/shakespeares-handwriting-in-the-book-of-sir-thomas-more.

5. William Shakespeare, *The Tempest,* http://shakespeare.mit.edu/tempest/full.html.

6. *Internet Encyclopedia of Philosophy,* "Gottfried Leibniz: Metaphysics," http://www.iep.utm.edu/leib-met/.

7. Jonathan Swift, *Gulliver's Travels* (New York: Signet Classic, 1983).

8. Voltaire, *Candide; or, Optimism,* trans. Peter Constantine (New York: Modern Library, 2005).

9. "The Mayflower Compact," 1620, http://mayflowerhistory.com/mayflower-compact/.

10. John Winthrop, "City Upon a Hill" sermon, 1630, http://www.digitalhistory.uh.edu/disp_textbook.cfm?smtID=3&psid=3918.

11. John M. Barry, "God, Government, and Roger Williams' Big Idea," *Smithsonian,* January 2012, http://www.smithsonianmag.com/history/god-government-and-roger-williams-big-idea-6291280/.

12. "Frame of Government of Pennsylvania," May 5, 1682, http://avalon.law.yale.edu/17th_century/pa04.asp.

13. Declaration of Independence, http://www.ushistory.org/declaration/document/.

14. John Locke, *Two Treatises on Government* (London, 1821), Bartleby.com, 2010, http://www.bartleby.com/169/.

15. Aristotle, *Nicomachean Ethics,* http://classics.mit.edu/Aristotle/nicomachaen.html.

16. "Thomas Jefferson Quotes," https://www.goodreads.com/author/quotes/1673.Thomas_Jefferson.

17. William Faulkner, *Requiem for a Nun* (New York: Random House, 1951).

18. Mark Twain, *The Adventures of Huckleberry Finn* (New York: Random House, 1996).

19. Mark Twain, autobiographical dictation, September 13, 1907, in *Autobiography of Mark Twain,* vol. 3 (Berkeley: University of California Press, 2015).

20. @realDonaldTrump, February 17, 2017.

21. Ibid., February 4, 2017.

22. "World Service Global Poll: Negative Views of Russia on the Rise," *BBC,* April 6, 2014, http://www.bbc.co.uk/mediacentre/latestnews/2014/world-service-country-poll.

23. Eric Brown, "In Gallup Poll, the Biggest Threat to World Peace Is . . . America?" *International Business Times,* January 2, 2014, http://www.ibtimes.com/gallup-poll-biggest-threat-world-peace-america-1525008.

24. Aldous Huxley, *Brave New World* (New York and London: Harper & Brothers, 1946).

25. George Orwell, *1984* (London: Secker & Warburg, 1949).

26. Sinclair Lewis, *It Can't Happen Here* (New York: New American Library, 1970).

Chapter 4: How Could a Trump Triumph?

1. Jeremy Rifkin, "New Technology and the End of Jobs," Converge.org, http://www.converge.org.nz/pirm/nutech.htm.

2. Gwynn Guilford, "Everything We Thought We Knew About Free Trade Is Wrong," *Quartz,* 2016, https://qz.com/840973/everything-we-thought-we-knew-about-free-trade-is-wrong/.

3. Ben Casselman, "Manufacturing Jobs Are Never Coming Back," *FiveThirtyEight,* March 18, 2016, https://fivethirtyeight.com/features/manufacturing-jobs-are-never-coming-back/.

4. "Impacts of Technological Change on Productivity," Boundless.com, https://www.boundless.com/economics/textbooks/boundless-economics-textbook/economic-growth-20/productivity-98/impacts-of-technological-change-on-productivity-370-12467/.

5. Ibid.

6. Harold L. Sirkin, Michael Zinser, and Justin Rose, "The Robotics Revolution: The Next Great Leap in Manufacturing," bcg.perspectives, September 23, 2015, https://www.bcgperspectives.com/content/articles/lean-manufacturing-innovation -robotics-revolution-next-great-leap-manufacturing/.

7. Wolfgang Lehmacher, "Don't Blame China for Taking U.S. Jobs," *Fortune*, November 8, 2016, http://fortune.com/2016/11/08/china-automation-jobs/.

8. David Rotman, "How Technology Is Destroying Jobs," *MIT Technology Review,* June 12, 2013, https://www.technologyreview.com/s/515926/how-technology-is -destroying-jobs/.

9. Jesse Eisinger, "Trump's Treasury Secretary Pick Is a Lucky Man. Very Lucky," *ProPublica,* December 1, 2016, https://www.propublica.org/article/trumps-treasury -secretary-pick-steven-mnuchin-is-a-lucky-man.

10. David Bier, "President Trump's 6 Biggest Threats to Liberty," *Learn Liberty,* January 20, 2017, http://www.learnliberty.org/blog/president-trumps-6-biggest -threats-to-liberty/.

11. John Bohannon, "Scientists to Trump: Torture Doesn't Work," *Science,* January 27, 2017, http://www.sciencemag.org/news/2017/01/scientists-trump-torture -doesn-t-work.

12. "Full Text: Donald Trump Announces a Presidential Bid," *Washington Post,* June 16, 2015, https://www.washingtonpost.com/news/post-politics/wp/2015/06/16/full -text-donald-trump-announces-a-presidential-bid/?utm_term=.c465ed27cfbb.

13. Alex Park, "These Charts Show How Ronald Reagan Actually Expanded the Federal Government," *Mother Jones,* December 30, 2014, http://www.motherjones .com/mojo/2014/12/ronald-reagan-big-government-legacy.

14. Allen Frances, "A Debate on the Pros and Cons of Aging and Death," *Huffington Post*, December 24, 2016, http://www.huffingtonpost.com/allen-frances/a-debate -on-the-pros-and-_b_13843296.html.

15. Michael Kruse,"The 199 Most Donald Trump Things Donald Trump Has Ever Said," *Politico,* August 14, 2015, http://www.politico.com/magazine/story/2015/08 /the-absolute-trumpest-121328.

16. Theodor W. Adorno et al., *The Authoritarian Personality* (New York: Harper, 1950).

17. Matthew MacWilliams, "The One Weird Trait That Predicts Whether You're a Trump Supporter," *Politico,* January 17, 2016, http://www.politico.com/magazine /story/2016/01/donald-trump-2016-authoritarian-213533.

18. Bobby Azarian, "A Neuroscientist Explains What May Be Wrong with Trump Supporters' Brains," *Raw Story,* August 4, 2016, http://www.rawstory.com/2016/08 /a-neuroscientist-explains-what-may-be-wrong-with-trump-supporters-brains/.

19. Peter Beinart, "Fear of a Female President," *Atlantic,* October 16, 2016, http://www .theatlantic.com/magazine/archive/2016/10/fear-of-a-female-president/497564/.

20. Abigail Geiger, "Number of Women Leaders Around the World Has Grown, but They're Still a Small Group," Pew Research Center, March 8, 2017, http:// pewresearch.org/fact-tank/2017/03/08/women-leaders-around-the-world/.

21. Michael D'Antonio, "Is Donald Trump Racist? Here's What the Record Shows," *Fortune,* June 7, 2016, http://fortune.com/2016/06/07/donald-trump-racism -quotes/.

22. Jacob Bogage, "Whom Are You Voting For? This Guy Can Read Your Mind," *Washington Post,* June 23, 2016, https://www.washingtonpost.com/news/the-switch/wp/2016/06/23/whom-are-you-voting-for-this-guy-can-read-your-mind/?utm_term=.4320db533013.

23. Jason Le Miere, "Did the Media Help Donald Trump Win? $5 Billion in Free Advertising Given to President-Elect," *International Business Times,* November 9, 2016, http://www.ibtimes.com/did-media-help-donald-trump-win-5-billion-free-advertising-given-president-elect-2444115.

24. Lorraine Boissoneault, "How the 19th-Century Know Nothing Party Reshaped American Politics," *Smithsonian.com,* January 26, 2017, http://www.smithsonianmag.com/history/immigrants-conspiracies-and-secret-society-launched-american-nativism-180961915/.

25. Tina Irvine, "The Striking Parallels Between Trump's Xenophobia and the Americanization Movement of the 1910s," *OMNIA,* September 8, 2016, https://omnia.sas.upenn.edu/story/striking-parallels-between-trump%E2%80%99s-xenophobia-and-americanization-movement-1910s.

26. Jane Mayer, *Dark Money: The Hidden History of the Billionaires Behind the Rise of the Radical Right* (New York: Doubleday, 2016).

27. Jeff Nesbit, *Poison Tea: How Big Oil and Big Tobacco Invented the Tea Party and Captured the GOP* (New York: St. Martin's Press, 2016).

28. Robert Reich, "Why the Republican's Old Divide-and-Conquer Strategy—Setting Working Class Against the Poor—Is Backfiring," *RobertReich.org,* January 9, 2014, http://robertreich.org/post/72770488951.

29. Seth J. Hill and Chris Tausanovitch, "A Disconnect in Representation? Comparison of Trends in Congressional and Public Polarization," *Journal of Politics* 77, no. 4 (October 2015), http://www.journals.uchicago.edu/doi/abs/10.1086/682398.

30. Ilyana Kuziemko and Ebonya Washington, "Why Did the Democrats Lose the South? Bringing New Data to an Old Debate," National Bureau of Economic Research, November 2015, http://www.nber.org/papers/w21703.

31. "The Polarization of the Congressional Parties," Voteview.com, March 21, 2015, http://www.voteview.com/political_polarization_2014.htm.

32. "Partisanship and Political Animosity in 2016," Pew Research Center, June 22, 2016, http://www.people-press.org/2016/06/22/partisanship-and-political-animosity-in-2016/.

33. Carroll Doherty, "7 Things to Know About Polarization in America," Pew Research Center, June 12, 2014, http://www.pewresearch.org/fact-tank/2014/06/12/7-things-to-know-about-polarization-in-america/.

34. Jennifer Agiesta, "CNN/ORC Poll: A Nation Divided, and Is It Ever," *CNN,* November 27, 2016, http://www.cnn.com/2016/11/27/politics/cnn-poll-division-donald-trump/.

35. Roger Sollenberger, "The Electoral College Is Pointless and Unfair, and Has Been That Way for 200 Years," *Paste,* December 16, 2016, https://www.pastemagazine.com/articles/2016/12/the-electoral-college-is-pointless-and-unfair-and.html.

Chapter 5: Trump, Tribalism, and the Attack on Democracy

1. Mike Godwin, "Sure, Call Trump a Nazi. Just Make Sure You Know What You're Talking About," *Washington Post,* December 14, 2015, https://www.washingtonpost

.com/amphtml/posteverything/wp/2015/12/14/sure-call-trump-a-nazi-just-make
-sure-you-know-what-youre-talking-about/.

2. Erica Chenoweth, "How Social Media Helps Dictators," *Foreign Policy,* November 16, 2016, http://foreignpolicy.com/2016/11/16/how-social-media-helps-dictators/.

3. Maria Konnikova, "Revisiting Robbers Cave: The Easy Spontaneity of Intergroup Conflict," *Scientific American,* September 5, 2012, https://blogs.scientificamerican.com/literally-psyched/revisiting-the-robbers-cave-the-easy-spontaneity-of-intergroup -conflict/.

4. Maureen B. Costello, *The Trump Effect: The Impact of the Presidential Campaign on Our Nation's Schools,* Southern Poverty Law Center, https://www.splcenter.org/sites/default/files/splc_the_trump_effect.pdf.

5. Jim Goad, "The Progressive Glossary," *Taki's Magazine,* May 20, 2013, http://takimag.com/article/the_progressive_glossary_jim_goad/print#axzz4.

Chapter 6: Defending Democracy: The Path Forward

1. "Lux: Why the Darkness Will Pass and Progressive Populism Will Light the Way Forward," *Democratic Strategist,* December 14, 2016, http://thedemocraticstrategist .org/2016/12/lux-why-the-darkness-will-pass-and-progressive-populism-will -light-the-way-forward/.

2. Saul David Alinsky, *Rules for Radicals: A Pragmatic Primer for Realistic Radicals* (New York: Vintage Books, 1971).

3. Bonnie Azab Powell, "Framing the Issues: UC Berkeley Professor George Lakoff Tells How Conservatives Use Language to Dominate Politics," *UC Berkeley News,* October 27, 2003, https://www.berkeley.edu/news/media/releases/2003/10/27 _lakoff.shtml.

4. Robert Cruickshank, "Why Aren't Progressives as Good at Politics as Conservatives?" *Daily Kos,* May 22, 2011, http://www.dailykos.com/story/2011/5/22/978274/-.

5. "A Short History," MoveOn.org, https://front.moveon.org/a-short-history/.

6. "Indivisible: A Practical Guide for Resisting the Trump Agenda," https://www .indivisibleguide.com/.

7. SwingLeft website, https://swingleft.org/.

8. Liz Kennedy, "Voter Suppression Laws Cost Americans Their Voices at the Polls," Center for American Progress, November 11, 2017, https://www.americanprogress .org/issues/democracy/reports/2016/11/11/292322/voter-suppression-laws-cost -americans-their-voices-at-the-polls/.

9. Emma Green, "These Conservative Christians Are Opposed to Trump—and Suffering the Consequences," *Atlantic,* February 11, 2017, https://www.theatlantic.com/politics/archive/2017/02/conservative-christians-disagreement-trump/516132/.

10. Ronald E. Riggio, "Narcissism and the U.S. Presidency," *Psychology Today,* October 2016, https://www.psychologytoday.com/blog/cutting-edge-leadership /201610/narcissism-and-the-us-presidency.

11. Jimmy Carter, "Energy and the National Goals—A Crisis of Confidence," *AmericanRhetoric.com,* http://americanrhetoric.com/speeches/jimmycartercrisisof confidence.htm.

12. Accept climate change is real and do something about it (65 percent of Americans). "New Poll: Most Americans Want Government to Combat Climate Change,

but Voters Deeply Divided Along Party Lines on Paying for Solutions," Associated Press–NORC Center for Public Affairs Research, Sept 14, 2016, http://www.apnorc.org/PDFs/EnergyClimate/Press%20Release_EPIC%20AP-NORC%20Energy%20Policy%20Poll_Final.pdf.

13. Emphasize alternative energy over oil and gas (84 percent of registered voters). Evan Lehmann, "Many More Republicans Now Believe in Climate Change," *Scientific American,* April 27, 2016, https://www.scientificamerican.com/article/many-more-republicans-now-believe-in-climate-change/.

14. Raise taxes on the wealthy (63 percent of Americans) and on corporations (67 percent of Americans). Frank Newport, "Majority Say Wealthy Americans, Corporations Taxed Too Little," Gallup, April 18, 2017, http://www.gallup.com/poll/208685/majority-say-wealthy-americans-corporations-taxed-little.aspx.

15. Follow policies that promote a more equal distribution of wealth (63 percent of Americans). Frank Newport, "Americans Continue to Say U.S. Wealth Distribution Is Unfair," Gallup, May 4, 2015; http://www.gallup.com/poll/182987/americans-continue-say-wealth-distribution-unfair.aspx.

16. Protect Social Security (over 80 percent of Americans). "The Importance of Economic Issues," Associated Press–NORC Center for Public Affairs Research, February 2016, http://www.apnorc.org/projects/Pages/the-importance-of-economic-issues.aspx.

17. Government has the responsibility to ensure health coverage for all (58 percent of Americans). Jocelyn Kiley, "Public Support for 'Single Payer' Health Coverage Grows, Driven by Democrats," Pew Research Center, June 23, 2017, http://www.pewresearch.org/fact-tank/2017/06/23/public-support-for-single-payer-health-coverage-grows-driven-by-democrats/.

18. Preserve Medicare (77 percent). Mire Norton, Bianco DiJulio, and Mollyann Brody, "Medicare and Medicaid at 50," Kaiser Family Foundation, July 17, 2015, http://www.kff.org/medicaid/poll-finding/medicare-and-medicaid-at-50/.

19. Negotiate lower prices for prescription drugs (70 percent of Americans). David Nather, "STAT-Harvard Poll: Dismayed by Drug Prices, Public Supports Democrats' Ideas," Harvard University, November 2015, http://www.harvard.edu/media-relations/stat-harvard-poll-dismayed-by-drug-prices-public-supports-democrats-ideas.

20. Reduce government waste and deficits (77 percent of Americans). Brian Montopoli, "Poll: Americans Split on What to Cut from Government," *CBS News,* January 14, 2011, http://www.cbsnews.com/news/poll-americans-split-on-what-to-cut-from-government/.

21. Campaign finance reform (85 percent of Americans). Nicholas Confessore and Megan Thee-Brenan, "Poll Shows Americans Favor an Overhaul of Campaign Financing," *New York Times,* June 2, 2015, https://www.nytimes.com/2015/06/03/us/politics/poll-shows-americans-favor-overhaul-of-campaign-financing.html.

22. Congress is too partisan and gridlocked (78 percent). Lydia Saad, "Gridlock Is Top Reason Americans Are Critical of Congress," Gallup, June 12, 2013, http://www.gallup.com/poll/163031/gridlock-top-reason-americans-critical-congress.aspx.

23. Improving the educational system is one of the public's top policy priorities (67 percent of Americans). Pew Research Center, "Public's Policy Priorities Reflect Changing Conditions at Home and Abroad," January 15, 2015, http://www

.people-press.org/2015/01/15/publics-policy-priorities-reflect-changing-conditions
-at-home-and-abroad/

24. Raise the minimum wage (73 percent of Americans). Bruce Drake, "Polls Show
Strong Support for Minimum Wage Hike," Pew Research Center, March 2, 2014,
http://www.pewresearch.org/fact-tank/2014/03/04/polls-show-strong-support
-for-minimum-wage-hike/.

25. Put people to work on urgent infrastructure repairs (64 percent of Ameri-
cans). Frank Newport, "Americans Say 'Yes' to Spending More on VA, Infrastruc-
ture," Gallup, March 21, 2016, http://www.gallup.com/poll/190136/ameri-%20
cans-say-yes-spending-infrastructure.aspx.

26. Lower tax rates for businesses and manufacturers that create jobs in the United
States (79 percent of Americans) and enact a federal jobs creation law that would spend
government money for a program designed to create more than one million new jobs
(72 percent of Americans). Jeffrey M. Jones, "Americans Widely Back Government Job
Creation Proposals," Gallup, March 20, 2013, http://www.gallup.com/poll/161438
/americans-widely-back-government-job-creation-proposals.aspx.

27. Establish stricter policies to prevent people from overstaying their visas (77 per-
cent); allow those born in the United States to illegal immigrants to remain here (72
percent); establish a way for most immigrants currently here illegally to stay legally
(62 percent); keep unqualified illegal immigrants from receiving government ben-
efits (73 percent); don't build a wall between us and Mexico (73 percent); take in
carefully vetted civilian refugees escaping violence and war (61 percent); encourage
more highly skilled people from around the world to immigrate to the United States
to work (58 percent). Bob Suls, "Less Than Half the Public Views Border Wall as
an Important Goal for U.S. Immigration Policy," Pew Research Center, January 6,
2017, http://www.pewresearch.org/fact-tank/2017/01/06/less-than-half-the-public
-views-border-wall-as-an-important-goal-for-u-s-immigration-policy/.

28. Lower health care costs (67 percent of Americans). Ashley Kirzinger, Bryan
Wu, and Mollyann Brodie, "Kaiser Health Tracking Poll: Health Care Priorities
for 2017," Kaiser Family Foundation, January 6, 2017, http://www.kff.org/health
-costs/poll-finding/kaiser-health-tracking-poll-health-care-priorities-for-2017/.

29. Support continued federal funding for Planned Parenthood (60 percent of
Americans). "Majority Says Any Budget Deal Must Include Planned Parenthood
Funding," Pew Research Center, September 28, 2015, http://www.people-press
.org/2015/09/28/majority-says-any-budget-deal-must-include-planned-parenthood
-funding/.

30. Keep abortion legal (59 percent of Americans). Michael Lipka and John Gram-
lich, "5 Facts About Abortion," Pew Research Center, January 26, 2017, http://
www.pewresearch.org/fact-tank/2017/01/26/5-facts-about-abortion/.

31. Reduce the prison population (69 percent of Americans). "ACLU Nationwide
Poll on Criminal Justice Reform," ACLU, July 15, 2015, https://www.aclu.org
/other/aclu-nationwide-poll-criminal-justice-reform.

32. Decriminalize drug addiction and mental illness ((87 percent of Americans).
"ACLU Nationwide Poll on Criminal Justice Reform," ACLU, July 15, 2015,
https://www.aclu.org/other/aclu-nationwide-poll-criminal-justice-reform.

33. Provide adequate treatment for drug addiction and mental health (67 percent of
Americans). "America's New Drug Policy Landscape," Pew Research Center, April 2,
2014, http://www.people-press.org/2014/04/02/americas-new-drug-policy-landscape/.

34. Douglas Keene, "Solution Aversion: Denying Problems When We Don't Like the Solutions," *Jury Room,* January 9, 2015, http://keenetrial.com/blog/2015/01/09 /solution-aversion-denying-problems-when-we-dont-like-the-solutions/.

Chapter 7: Sustaining Our Brave New World

1. "World Population Clock," Worldometers, http://www.worldometers.info/world -population/.

2. Max Roser and Esteban Ortiz-Ospina, "Global Extreme Poverty," OurWorld InData.org, 2016, https://ourworldindata.org/world-poverty/.

3. Max Roser, "Fertility," OurWorldInData.org, 2016, https://ourworldindata.org /fertility/.

4. "Country Comparison: Birth Rate," *The World Factbook,* Central Intelligence Agency, 2016, https://www.cia.gov/library/publications/the-world-factbook/rank order/2054rank.html.

5. "China Experiencing Baby Boom Now That One-Child Rule Is Lifted," *South China Morning Post,* December 29, 2016, http://www.scmp.com/news/china/policies -politics/article/2057945/china-experiencing-baby-boom-now-one-child-rule-lifted.

6. David Canning, "The Causes and Consequences of the Demographic Transition," (working paper, Harvard School of Public Health, July 2011), http://citeseerx .ist.psu.edu/viewdoc/download?doi=10.1.1.698.4763&rep=rep1&type=pdf.

7. Kim Parker, "Parenthood and Happiness: It's More Complicated Than You Think," Pew Research Center, February 7, 2014, http://www.pewresearch.org /fact-tank/2014/02/07/parenthood-and-happiness-its-more-complicated-than -you-think/.

8. "World Population Projected to Reach 9.7 Billion by 2050," United Nations Department of Economic and Social Affairs, July 29, 2019, http://www.un.org/en /development/desa/news/population/2015-report.html.

9. F. Brinley Bruton, "Turkey's President Erdogan Calls Women Who Work 'Half Persons,'" *NBC News,* June 8, 2016, http://www.nbcnews.com/news/world/turkey -s-president-erdogan-calls-women-who-work-half-persons-n586421.

10. "Zika Virus and Complications: Questions and Answers," World Health Organization, updated November 15, 2016, http://www.who.int/features/qa/zika/en/.

11. "Family Planning Strategy Overview," Gates Foundation, http://www.gates foundation.org/What-We-Do/Global-Development/Family-Planning.

12. John Michael Greer, "How Civilizations Fall: A Theory of Catabolic Collapse," 2005, http://ecoshock.org/transcripts/greer_on_collapse.pdf.

13. "Household Final Consumption Expenditure, etc. (% of GDP)," World Bank, 2016, http://data.worldbank.org/indicator/NE.CON.PETC.ZS.

14. Tim Worstall, "Keynes' 15 Hour Work Week Is Here Right Now," *Forbes,* October 16, 2015, https://www.forbes.com/sites/timworstall/2015/10/16/keynes -15-hour-work-week-is-here-right-now/.

15. Herbert Marcuse, *One-Dimensional Man: Studies in the Ideology of Advanced Industrial Society* (Boston: Beacon Press, 1991).

16. Torbin M. Andersen et al., "The Nordic Model: Embracing Globalization and Sharing Risks," Research Institute of the Finnish Economy (ETLA), 2007, https:// www.etla.fi/en/publications/b232-en/.

17. Peter Dockrill, "Ants Respond as a Collective 'Superorganism' When They Sense a Predator," *Science Alert,* November 13, 2015, http://www.sciencealert.com/ants-respond-as-a-collective-superorganism-when-they-sense-a-predator.

18. E. O. Wilson, *Biophilia* (Cambridge, MA: Harvard University Press, 1984).

19. Linnie Marsh Wolfe, *John of the Mountains* (Madison: University of Wisconsin Press, 1979).

20. *Life After People,* TV series, directed by David de Vries, 2009–11, http://www.imdb.com/title/tt1433058/.

Chapter 8: The Pursuit of Happiness

1. Lucretius, *On Nature* (Indianapolis: Bobbs-Merrill, 1965).

2. Epicurus, Letter to Menoeceus, http://www.epicurus.net/en/menoeceus.html.

3. "Stoicism vs. Epicureanism," Academy of Ideas, http://academyofideas.com/2014/03/stoicism-vs-epicureanism/.

4. John Stuart Mill and Jeremy Bentham, *Utilitarianism and Other Essays* (Harmondsworth, England: Penguin Books, 1987).

5. Sigmund Freud, *Beyond the Pleasure Principle,* Bartleby.com, 2010, http://www.bartleby.com/276/.

6. Simon Moss, "Set Point Theory," Sico Tests, July 17, 2016, http://www.sicotests.com/psyarticle.asp?id=399.

7. Sonja Lyubomirsky, "Hedonic Adaptation to Positive and Negative Experiences," in *The Oxford Handbook of Stress, Health, and Coping,* ed. Susan Folkman (Oxford: Oxford University Press, 2011).

8. Jennifer Welsh, "Happiness Is U-Shaped: It Drops in Middle Age, Rises Later," *Live Science,* April 19, 2011, http://www.livescience.com/13788-happiness-life time.html.

9. Carol Graham and Soumya Chattopadhyay, "Gender and Well-Being Around the World: Some Insights from the Economics of Happiness," (working paper, Human Capital and Economic Opportunity: A Global Working Group, University of Chicago, May 2012), http://humcap.uchicago.edu/RePEc/hka/wpaper/Graham_Chattopadhyay_2012_GenderandWellBeing.pdf.

10. Brianna L. Kirkpatrick, "Personality and Happiness" (honors thesis, University of San Diego, 2015), http://digital.sandiego.edu/cgi/viewcontent.cgi?article=1013&context=honors_theses.

11. J. Helliwell, R. Layard, and J. Sachs, eds., *World Happiness Report 2016,* vol. 1, update (New York: Sustainable Development Solutions Network, 2016), http://worldhappiness.report/ed/2016/.

12. "Bhutan's Gross National Happiness Index," Oxford Poverty & Human Development Initiative, http://www.ophi.org.uk/policy/national-policy/gross-national-happiness-index/.

13. Richard Florida, "Income Inequality Leads to Less Happy People," *Citylab,* December 21, 2015, http://www.citylab.com/politics/2015/12/income-inequality-makes-people-unhappy/416268/.

14. Keith Breene, "The World's Happiest Countries in 2016," World Economic Forum, November 14, 2016, https://www.weforum.org/agenda/2016/11/the-worlds-happiest-countries-in-2016/.

15. George Herbert, *Jacula Prudentum; or, Outlandish Proverbs, Sentences, &C.* (London: T. Maxey for T. Garthwait, 1651).

16. Edmund Burke, *Letters on a Regicide Peace,* vol. 3 of *Select Works of Edmund Burke,* (Indianapolis: Liberty Fund, 1999), http://oll.libertyfund.org/titles/burke-select-works-of-edmund-burke-vol-3.

17. Elizabeth W. Dunn and Michael Norton, "Don't Indulge. Be Happy," *New York Times,* July 7, 2012, http://www.nytimes.com/2012/07/08/opinion/sunday/dont-indulge-be-happy.html.

18. Betsey Stevenson and Justin Wolfers, "Subjective Well-Being and Income: Is There Any Evidence of Satiation?," Brookings Institution, April 29, 2013, https://www.brookings.edu/research/subjective-well%E2%80%90being-and-income-is-there-any-evidence-of-satiation/.

19. Richard M. Ryan and Edward L. Deci, "On Happiness and Human Potentials: A Review of Research on Hedonic and Eudaimonic Well-Being," *Annual Review of Psychology* 52 (February 2001): 141–66, http://www.annualreviews.org/doi/abs/10.1146/annurev.psych.52.1.141.

20. Debra Umberson and Jennifer Karas Montez, "Social Relationships and Health: A Flashpoint for Health Policy," *Journal of Health and Social Behavior* 51, suppl (2010): S54–S66, https://www.ncbi.nlm.nih.gov/pmc/articles/PMC3150158/#!po=0.349650.

21. R. C. Rosen et al., "Sexual Well-Being, Happiness, and Satisfaction, in Women: The Case for a New Conceptual Paradigm," *Journal of Sex and Marital Therapy* 34, no. 4 (2008): 291–97, https://www.ncbi.nlm.nih.gov/labs/articles/18576229/.

22. Aristotle, *The Nicomachean Ethics,* ed. W. D. Ross and Lesley Brown (Oxford: Oxford University Press, 2009).

23. Natasha Gilbert, "Altruism Can Be Explained by Natural Selection," *Nature,* August 25, 2010, http://www.nature.com/news/2010/100825/full/news.2010.427.html.

24. Maria Konnikova, "The Psychology Behind Gift-Giving and Generosity," *Scientific American,* January 4, 2012, https://blogs.scientificamerican.com/literally-psyched/the-psychology-behind-gift-giving-and-generosity/.

25. Ojibwa, "The Potlatch," *Native American Netroots,* August 13, 2010, http://nativeamericannetroots.net/diary/631.

26. A. Krosnick, "The Diabetes and Obesity Epidemic Among the Pima Indians," *New Jersey Medicine* 97, no. 8 (2000): 31–37, https://www.ncbi.nlm.nih.gov/pubmed/10959174.

27. "Mental and Substance Use Disorders," Substance Abuse and Mental Health Services Administration, last modified March 8, 2016, https://www.samhsa.gov/disorders.

28. Aldous Huxley, *Brave New World* (New York and London: Harper & Brothers, 1946).

29. Aldous Huxley, *Drugs That Shape Men's Minds* (New York: Knopf, 1962).

30. F. R. Ervin et al., "Voluntary Consumption of Beverage Alcohol by Vervet Monkeys: Population Screening, Descriptive Behavior and Biochemical Measures," *Pharmacology Biochemistry and Behavior* 36, no. 2 (June 1990): 367–73, https://www.ncbi.nlm.nih.gov/pubmed/2356209.

31. Sarah Tse, "Natural High: Animals That Use Drugs in the Wild," *The Science Explorer,* October 26, 2015, http://thescienceexplorer.com/nature/natural-high-animals-use-drugs-wild.

32. *Wikipedia,* "Effect of Psychoactive Drugs on Animals," https://en.wikipedia.org/wiki/Effect_of_psychoactive_drugs_on_animals.

33. Edith V. Sullivan, R. Adron Harris, and Adolf Pfefferbaum, "Alcohol's Effects on Brain and Behavior," *Alcohol Research & Health* 33, no. 1–2 (2010): 127–43, https://www.ncbi.nlm.nih.gov/pmc/articles/PMC3625995/.

34. "CDC Guideline for Prescribing Opioids for Chronic Pain—United States, 2016," *Morbidity and Mortality Weekly Report,* posted March 15, 2016, https://www.cdc.gov/mmwr/volumes/65/rr/rr6501e1.htm.

35. Allen Frances, "Opioid Companies Lobby Against Medical Marijuana," *Huffington Post,* September 12, 2016, http://www.huffingtonpost.com/allen-frances/opioid-companies-lobby-ag_b_11287182.html.

36. Alan Frazer and Julie G. Hensler, "Serotonin Involvement in Physiological Function and Behavior," *NCBI,* 1999, https://www.ncbi.nlm.nih.gov/books/NBK27940.

37. "Drug Trafficking by the Numbers," The Recovery Village, https://www.therecoveryvillage.com/drug-addiction/drug-trafficking-by-the-numbers/.

Chapter 9: Team Earth

1. Eillie Anzilotti, "Cities Have Been Trying to Curb Dog Poop for Centuries," *Citylab,* May 3, 2016, http://www.citylab.com/navigator/2016/05/a-brief-history-of-dog-poop-etiquette-campaigns/480870/.

2. Bradford Plumer, "The Origins of Anti-Litter Campaigns," *Mother Jones,* May 22, 2006, http://www.motherjones.com/mojo/2006/05/origins-anti-litter-campaigns.

3. Natalie Angier, *Woman: An Intimate Geography* (Boston: Houghton Mifflin, 1999).

4. John Moore Williams, "The Hotly Contested History of the Seat Belt," *Esurance* (blog), http://blog.esurance.com/seat-belt-history/.

5. Committee on Secondhand Smoke Exposure and Acute Coronary Events and Board on Population Health and Public Health Practice, *Secondhand Smoke Exposure and Cardiovascular Effects: Making Sense of the Evidence* (Washington, DC: National Academies Press, 2010), p. 109, https://www.nap.edu/read/12649/chapter/7.

6. "The Environmental Movement's Greatest Success Story: Ozone Layer Begins to Heal," United Nations Environment Programme, July 29, 2016, http://www.unep.org/stories/story/environmental-movement%E2%80%99-greatest-success-story-ozone-layer-begins-heal.

7. Richard Kerr, "Acid Rain Control: Success on the Cheap," *Science,* November 6, 1998, http://science.sciencemag.org/content/282/5391/1024.full?ck=ck.

8. Zachary Davies Boren, "China's Coal Bubble: 210 New Coal-Fired Power Plants Were Approved in 2015," Greenpeace, March 2, 2016, http://energydesk.greenpeace.org/2016/03/02/china-coal-bubble-210-power-plants/.

9. Bonnie Eisenberg and Mary Ruthsdotter, "History of the Women's Rights Movement," National Women's History Project, 1988, http://www.nwhp.org/resources/womens-rights-movement/history-of-the-womens-rights-movement/.

10. Borgna Brunner, "Timeline: Key Moments in Black History," Infoplease.com, http://www.infoplease.com/spot/bhmtimeline.html.

11. "The American Gay Rights Movement: A Timeline," Infoplease.com, http://www.infoplease.com/ipa/A0761909.html.

12. John Breuilly, ed., *The Oxford Handbook of the History of Nationalism* (Oxford: Oxford University Press, 2013).

Index

BOOKS BY
ALLEN FRANCES

SAVING NORMAL
An Insider's Revolt against Out-of-Control Psychiatric Diagnosis, DSM-5, Big Pharma, and the Medicalization of Ordinary Life

A deeply fascinating and urgently important critique of the widespread medicalization of normality. Frances cautions that the newest edition of the "bible of psychiatry," the *Diagnostic and Statistical Manual of Mental Disorders-5 (DSM-5)*, is turning our current diagnostic inflation into hyperinflation by converting millions of "normal" people into "mental patients." *Saving Normal* is a call to all of us to reclaim the full measure of our humanity.

TWILIGHT OF AMERICAN SANITY
A Psychiatrist Analyzes the Age of Trump

A landmark book, from "one of the world's most prominent psychiatrists" (*The Atlantic*): Allen Frances analyzes the nation, viewing the rise of Donald J. Trump as darkly symptomatic of a deeper societal distress that must be understood if we are to move forward. Frances argues that Trump is "bad, not mad"—and that the real question to wrestle with is how we as a country could have chosen him as our leader. *Twilight of American Sanity* is an essential work for understanding our national crisis.